Tight Junctions in Inflammation and Cancer

Ajaz A. Bhat • Mohammad Haris •
Muzafar A. Macha • Punita Dhawan
Editors

Tight Junctions in Inflammation and Cancer

Springer

Editors
Ajaz A. Bhat
Department of Human Genetics-Precision Medicine in Diabetes, Obesity and Cancer Research Program
Sidra Medicine
Doha, Qatar

Mohammad Haris
Center for Advanced Metabolic Imaging in Precision Medicine, Department of Radiology, Perelman School of Medicine
University of Pennsylvania
Philadelphia, PA, USA

Muzafar A. Macha
Watson-Crick Centre for Molecular Medicine, Islamic University of Science and Technology
Jammu and Kashmir, India

Punita Dhawan
Department of Biochemistry and Molecular Biology
University of Nebraska Medical Center
Omaha, NE, USA

ISBN 978-981-99-2414-1 ISBN 978-981-99-2415-8 (eBook)
https://doi.org/10.1007/978-981-99-2415-8

This Springer imprint is published by the registered company Springer Nature Singapore Pte Ltd.
The registered company address is: 152 Beach Road, #21-01/04 Gateway East, Singapore 189721, Singapore

Preface

Cancer remains one of the most devastating diseases that plagues humanity, accounting for millions of deaths worldwide each year. Despite advances in diagnostics, treatments, and our understanding of the disease, cancer continues to pose significant challenges in both basic and clinical research. In recent years, the complex and multifaceted relationship between inflammation and cancer has garnered significant attention from researchers and clinicians worldwide. As our understanding of the underlying mechanisms and pathways evolves, it becomes increasingly important to consolidate and synthesize the vast body of knowledge that has been generated. This edited volume, *Tight Junctions in Inflammation and Cancer*, is an ambitious attempt to provide a comprehensive and up-to-date resource that delves into the intricate world of tight junctions and their role in the inflammation-cancer nexus.

Tight junctions, the integral components of cellular adhesion complexes, play a crucial role in maintaining the integrity and functionality of epithelial and endothelial barriers. Disruptions in tight junctions are often associated with an array of pathologies, including inflammatory diseases and cancer. The chapters in this book highlight the latest research findings and provide a detailed understanding of how tight junctions contribute to the initiation, progression, and metastasis of cancer, as well as their potential as therapeutic targets.

The book begins with an overview of the molecular structure and function of tight junctions, emphasizing their regulatory role in maintaining cellular homeostasis. Following this introduction, the subsequent chapters delve into the emerging evidence on the role of tight junctions in inflammation and their contribution to the development of cancer. We have assembled contributions from leading experts across the globe, who have shared their insights on the complex interplay between inflammation, tight junctions, and cancer.

The central chapters of the book explore the molecular mechanisms that link tight junction dysfunction with inflammation and cancer, including the role of specific tight junction proteins in the activation of oncogenic signaling pathways, the involvement of the immune system, and the effects of the tumor microenvironment. Furthermore, the book discusses the significance of tight junctions in metastasis and the process of epithelial-mesenchymal transition, which are both key aspects of cancer progression.

In the latter part of the book, the focus shifts to the potential applications of tight junction research in the development of innovative therapeutic strategies for cancer treatment. This section highlights the promise of targeting tight junctions in the modulation of cancer-associated inflammation and as a means of overcoming drug resistance. The book concludes with a discussion of future research directions and the potential impact of tight junction research on personalized medicine and cancer prevention.

Tight Junctions in Inflammation and Cancer is an invaluable resource for scientists, clinicians, and students interested in understanding the complex interplay between inflammation and cancer, with a specific focus on the role of tight junctions. We believe this book will not only serve as a reference guide for experts in the field but also inspire and facilitate the development of novel therapeutic strategies in the fight against cancer.

We would like to express our gratitude to all the authors who have contributed their expertise and insights to this volume, as well as the reviewers who have ensured the quality and rigor of the content. We hope that this book will stimulate further research, promote interdisciplinary collaboration, and ultimately contribute to our collective efforts to unravel the mysteries of inflammation and cancer.

Doha, Qatar — Ajaz A. Bhat
Philadelphia, PA, USA — Mohammad Haris
Jammu and Kashmir, India — Muzafar A. Macha
Omaha, NE, USA — Punita Dhawan

Contents

Editors and Contributors

About the Editors

Ajaz A. Bhat, a staff scientist at Sidra Medicine, Qatar, has an extensive background in peptide-based vaccine development, deregulated signaling pathways, and drug targets for various cancers. After earning his Ph.D. from AIIMS, India, he undertook postdoctoral training at Vanderbilt University Medical School, USA, expanding his understanding of colorectal, gastric, and esophageal cancers. His high-impact research, published in prestigious journals like *Nature Communications*, *GUT*, *Oncogene*, *Cancer Research*, and *Molecular Cancer*, earned him recognition as an outstanding researcher. He has received awards from the Immunology Society and the American Association of Cancer Research, and he actively serves on several immunology and cancer societies' editorial and review boards.

Mohammad Haris is an associate professor at the Center for Advanced Metabolic Imaging in Precision Medicine, Department of Radiology, Perelman School of Medicine, University of Pennsylvania, Philadelphia, USA. He leads the molecular and metabolic imaging research program to develop novel MR imaging methods to image and diagnose various human diseases at their incipient stage. His additional role provides interdisciplinary training projects and an educational environment on biomedical imaging to undergraduate, master's, and graduate students, emphasizing translational and clinical applications. He serves as a section editor of the *Journal of Translational Medicine*, managing editor of *Translational Medicine Communications*, and reviewer for many international scientific journals. He also holds the adjunct Professor position at Qatar University. After completing his Ph.D. in biomedical imaging from a reputed institution in India, Dr. Haris joined the Perelman School of Medicine at the University of Pennsylvania, where his focus was developing novel molecular and metabolic imaging techniques using NMR and MRI. Dr. Haris has published his research in leading journals, including *Nature Medicine*, *Nature Biomedical Engineering*, *Cancer Research*, *Molecular Cancer*, and *Cell*.

Muzafar A. Macha, currently an assistant professor and Ramalingaswami Fellow at Watson-Crick Centre for Molecular Medicine, IUST, Kashmir, India, possesses a strong background in Biochemistry, Head and Neck Cancers (HNCs), and therapeutic modalities development. He previously served as a postdoctoral fellow and later as an Assistant Professor at the University of Nebraska Medical Centre (UNMC), leading an independent HNC research group. His notable contributions, recognized with multiple awards, include the Young Investigator Award, Educational scholarship award, Scholar-In-Training Award, Best performance stipend award (2012, 2013), Gita Mittal Award/Gold Medal for the best basic Cancer Research in Ph.D. (AIIMS, New Delhi), and a gold medal for his master's degree (Jamia Hamdard, New Delhi). More recently, he received two prestigious Ramalingaswami & Ramanujan fellowships from the Govt. of India. He has also published his research in esteemed journals such as the *Journal of the National Cancer Institute*, *Oncogene*, *Carcinogenesis*, *Cancer Treatment Reviews*, *Advanced Drug Delivery Reviews*, *BBA-Molecular Basis of Disease*, *Oncotarget*, and *EBioMedicine*.

Punita Dhawan is currently working as a professor at the Department of Biochemistry and Molecular Biology, University of Nebraska Medical Center, Omaha, Nebraska, USA. Her research focuses on understanding novel biomarkers and therapeutic targets for colorectal cancer progression and metastasis. The field of epithelial barrier physiology with specific consideration of claudin proteins was born in 1998, and in 2005 she published a groundbreaking finding that, contrary to conventional wisdom, upregulated claudin-1 expression promotes sporadic colon cancer progression in JCI (J Clin Invest. 2005 Jul;115(7):1765–76). After that, her ongoing investigations to understand the molecular details of the role of this protein in regulating intestinal homeostasis and pathologies have resulted in publications in *Gastroenterology*, *GUT*, *Cancer Research*, *Mucosal Biology*, *Oncogene*, etc. In 2015, she demonstrated the role of Claudin-7 as a tumor suppressor in colorectal cancer, and this paper has received great attention. Her studies also focus on determining the role of cancer stem cells (CSCs) in colon cancer metastasis and chemoresistance. In this course, her group has generated novel reagents, genetically modified cell lines, and multiple mouse models with specific modifications of claudin-1 and claudin-7 proteins alone or in association with other significant co-founders but also established our leadership in this field of investigation. As PI or co-Investigator on federally funded grants, she laid the groundwork for the proposed research to perform these studies and have the expertise, leadership, and motivation necessary to carry out the proposed work successfully. In addition, a member of the iCaRe2 Biobank/Biorepository has access to a "state-of-the-art" patient cohort of CRC samples. As principal investigator or co-Investigator on federal, regional, and local grant funding, including NIH-R21, R01, and VA-merit awards aimed at intestinal pathologies and cancer, she effectively demonstrates leadership as the PI of this unique PI multidisciplinary grant proposal.

Contributors

Alanoud Abdulla Department of Human Genetics-Precision Medicine in Diabetes, Obesity and Cancer Program, Sidra Medicine, Doha, Qatar

Sabah Akhtar Translational Research Institute, Academic Health System, Hamad Medical Corporation, Doha, Qatar

Ammira Al-Shabeeb Akil Department of Human Genetics-Precision Medicine in Diabetes, Obesity and Cancer Program, Sidra Medicine, Doha, Qatar

Tayyiba Akbar Ali Division of Genomics and Translational Medicine, College of Health and Life Sciences, Hamad Bin Khalifa University, Doha, Qatar

Susmita Barman Department of Biochemistry and Molecular Biology, University of Nebraska Medical Center, Omaha, NE, USA

Ajaz A. Bhat Department of Human Genetics-Precision Medicine in Diabetes, Obesity and Cancer Research Program, Sidra Medicine, Doha, Qatar

Kabir H. Biswas Division of Biological and Biomedical Sciences, College of Health & Life Sciences, Hamad Bin Khalifa University, Doha, Qatar

Punita Dhawan Department of Biochemistry and Molecular Biology, University of Nebraska Medical Center, Omaha, NE, USA

Imadeldin Elfaki Department of Biochemistry, Prince Fahad Bin Sultan Chair for Biomedical Research, Faculty of Applied Medical Sciences, University of Tabuk, Tabuk, Saudi Arabia

Asma Fatima Division of Biological and Biomedical Sciences, College of Health & Life Sciences, Hamad Bin Khalifa University, Doha, Qatar

Iram Fatima Department of Biochemistry and Molecular Biology, University of Nebraska Medical Center, Omaha, NE, USA

Siri Chandana Gampa Department of Biotechnology, GITAM Institute of Science, GITAM (Deemed to be University), Visakhapatnam, India

Sireesha V. Garimella Department of Biotechnology, GITAM Institute of Science, GITAM (Deemed to be University), Visakhapatnam, India

Anupriya M. Geethakumari Division of Biological and Biomedical Sciences, College of Health & Life Sciences, Hamad Bin Khalifa University, Doha, Qatar

Saiprasad Gowrikumar Department of Biochemistry and Molecular Biology, University of Nebraska Medical Center, Omaha, NE, USA

Mohammad Haris Center for Advanced Metabolic Imaging in Precision Medicine, Department of Radiology, Perelman School of Medicine, University of Pennsylvania, Philadelphia, PA, USA

Sheema Hashem Department of Human Genetics, Sidra Medicine, Doha, Qatar

Rafia Jan Department of Pharmaceutical Sciences, University of Kashmir, Srinagar, Jammu and Kashmir, India

Inamu Rashid Khan Department of Zoology, School of Life Sciences, Central University of Kashmir, Ganderbal, Jammu and Kashmir, India

Tulaib Azam Khan Department of Biotechnology, School of Life Sciences, Central University of Kashmir, Ganderbal, Jammu and Kashmir, India

Sadaf Khursheed Department of Microbiology, Sher-I-Kashmir Institute of Medical Science (SKIMS), Soura, Jammu and Kashmir, India

Sana Khurshid Watson-Crick Centre for Molecular Medicine, Islamic University of Science and Technology, Awantipora, Jammu and Kashmir, India

Jayaprakash Narayana Kolla Institute of Molecular Genetics of the Czech Academy of Sciences, Prague, Czech Republic

Mudasir A. Kumar Watson-Crick Centre for Molecular Medicine, Islamic University of Science and Technology, Awantipora, Jammu and Kashmir, India

Muzafar A. Macha Watson-Crick Centre for Molecular Medicine, Islamic University of Science and Technology, Awantipora, Jammu and Kashmir, India

Sara Al Marzooqi Department of Human Genetics-Precision Medicine in Diabetes, Obesity and Cancer Program, Sidra Medicine, Doha, Qatar

Tariq Masoodi Human Immunology Department, Research Branch, Sidra Medicine, Doha, Qatar
Laboratory of Cancer Immunology and Genetics, Sidra Medicine, Doha, Qatar

Rashid Mir Department of Medical Lab Technology, Prince Fahad Bin Sultan Chair for Biomedical Research, Faculty of Applied Medical Sciences, University of Tabuk, Tabuk, Saudi Arabia

Roohi Mohi-ud-din Department of Pharmaceutical Sciences, University of Kashmir, Srinagar, Jammu and Kashmir, India

Magesh Muthu John D. Dingell VA Medical Center, Detroit, MI, USA
Department of Oncology, Barbara Ann Karmanos Cancer Institute, School of Medicine, Wayne State University, Detroit, MI, USA

Sabah Nisar St. Jude Children's Research Hospital, Memphis, TN, USA

Megan Lynn Otte Department of Biochemistry and Molecular Biology, University of Nebraska Medical Center, Omaha, NE, USA

Santhi Latha Pandrangi Department of Biochemistry and Bioinformatics, GITAM Institute of Science, GITAM (Deemed to be University), Visakhapatnam, India

Kristina Pravoverov Department of Biochemistry and Molecular Biology, University of Nebraska Medical Center, Omaha, NE, USA

Mark Primeaux Department of Biochemistry and Molecular Biology, University of Nebraska Medical Center, Omaha, NE, USA

Ravinder Reddy Center for Advanced Metabolic Imaging in Precision Medicine, Department of Radiology, Perelman School of Medicine, University of Pennsylvania, Philadelphia, PA, USA

Rahul Roy Centre for Biomedical Engineering, Indian Institute of Technology, New Delhi, India

Hana Sadida Department of Human Genetics-Precision Medicine in Diabetes, Obesity and Cancer Program, Sidra Medicine, Doha, Qatar

Amar Bahadur Singh Department of Biochemistry and Molecular Biology, University of Nebraska Medical Center, Omaha, NE, USA
VA Nebraska-Western Iowa Health Care System, Omaha, NE, USA
Buffet Cancer Center, University of Nebraska Medical Center, Omaha, NE, USA

Mayank Singh Department of Medical Oncology, Dr. BRAIRCH, All India Institute of Medical Sciences, New Delhi, India

Asfia Sultana Division of Biological and Biomedical Sciences, College of Health & Life Sciences, Hamad Bin Khalifa University, Doha, Qatar

Raju Lama Tamang Department of Biochemistry and Molecular Biology, University of Nebraska Medical Center, Omaha, NE, USA

S. M. Nasir Uddin Division of Biological and Biomedical Sciences, College of Health & Life Sciences, Hamad Bin Khalifa University, Doha, Qatar

Shahab Uddin Translational Research Institute, Academic Health System, Hamad Medical Corporation, Doha, Qatar

Burhan UlHaq Department of Zoology, School of Life Sciences, Central University of Kashmir, Ganderbal, Jammu and Kashmir, India

Santosh Kumar Yadav Department of Biochemistry and Molecular Biology, University of Nebraska Medical Center, Omaha, NE, USA

Inflammation and Cancer: Role of Tight Junctions

1

Kristina Pravoverov, Susmita Barman, Saiprasad Gowrikumar, Iram Fatima, Santosh Kumar Yadav, Megan Lynn Otte, Raju Lama Tamang, Mark Primeaux, Amar Bahadur Singh, and Punita Dhawan

Abstract

Tight junctions are critical for the homeostatic maintenance of epithelial barrier integrity, especially in the gastrointestinal tract. In addition, tight junction proteins serve as intermediary molecules in a variety of signaling pathways in both health and disease. Dysregulation of tight junction proteins as both a result of and secondary to infection, autoimmunity, or malignancy can either ameliorate or worsen disease, depending on the context. The intricacies of tissue- and pathology-specific contexts for barrier integrity and tight junction protein signaling are still poorly understood, but in this chapter, we will explore how tight junction proteins contribute to gut barrier integrity and how they are affected by inflammation during infection and gut dysbiosis and during inflammatory bowel diseases such as Crohn's disease and ulcerative colitis. Further, we will discuss the involvement of various tight junction proteins in major oncogenic signaling pathways and how their dysregulation in cancer promotes tumor proliferation, invasion, and metastasis. Then, we will look at a number of promising therapeutic strategies in development at the clinical or preclinical levels for targeting tight

K. Pravoverov · S. Barman · S. Gowrikumar · I. Fatima · S. K. Yadav · M. L. Otte · R. L. Tamang · M. Primeaux · P. Dhawan (✉)
Department of Biochemistry and Molecular Biology, University of Nebraska Medical Center, Omaha, NE, USA
e-mail: punita.dhawan@unmc.edu

A. B. Singh
Department of Biochemistry and Molecular Biology, University of Nebraska Medical Center, Omaha, NE, USA

VA Nebraska-Western Iowa Health Care System, Omaha, NE, USA

Buffet Cancer Center, University of Nebraska Medical Center, Omaha, NE, USA

A. A. Bhat et al. (eds.), *Tight Junctions in Inflammation and Cancer*, https://doi.org/10.1007/978-981-99-2415-8_1

junction proteins or the signaling pathways in which they are involved to treat infection and cancer.

Keywords

Cancer · Claudins · Inflammation · Oncogene · Tight junction · Proliferation · Metastasis

1.1 Introduction

Tight junctions (TJs) are critical for maintaining epithelial barrier function and integrity. They establish apical-basal polarity in epithelial cells and enable the selective paracellular diffusion of ions, solutes, and water across the barrier (Zihni et al. 2016; Muto et al. 2010; Al-Sadi et al. 2011). They are composed of transmembrane proteins including claudins, occludin, and junctional adhesion molecules (JAMs) that are anchored to underlying cytoplasmic plaque proteins known as zonula occludens (ZO)-1, -2, and -3 that connect to intracellular actin filaments (Rescigno 2011). In addition to their regulation of barrier function, TJ proteins are also involved in cell signaling pathways implicated in the pathogenesis and progression of various diseases (González-Mariscal et al. 2014; Pope et al. 2014a; Lu et al. 2013; Landy et al. 2016; Bhat et al. 2018; Roehlen et al. 2020; Wang et al. 2011; Zeisel et al. 2019; Zuo et al. 2020). In this chapter, we review the role of dysregulated TJs in inflammation and cancer, with a particular focus on the gastrointestinal system.

1.2 Barrier Dysfunction and Tight Junction Regulation During Inflammation

Recent studies have highlighted the importance of TJs in gastrointestinal epithelial barrier function in health and disease. It is common to find compromised intestinal barrier integrity in both gastrointestinal and systemic diseases, including infection, inflammatory bowel disease (IBD), and colitis-associated cancer (Fig. 1.1).

Claudins are the major determinants of epithelial permeability, and they can be broadly characterized as either barrier-forming/"tight" or pore-forming/"leaky," depending on their effect on membrane permeability (Günzel and Yu 2013). The structure of individual claudins determines their ability to form pores that vary in selectivity for solutes and water; as such, the changes in relative expression levels seen during inflammation can directly influence the pathology. Barrier-forming claudins such as claudin-1, -3, -4, -5, -6, -8, -12, -18, and -19 are highly selective and decrease membrane permeability (Furuse et al. 2002; Hou et al. 2010; Amasheh et al. 2005; Alexandre et al. 2007; Konrad et al. 2006). Pore-forming claudins such as claudin-2 and -15 enable increased paracellular permeability to ions and water, allowing substances to pass more freely across the epithelial barrier (Amasheh et al.

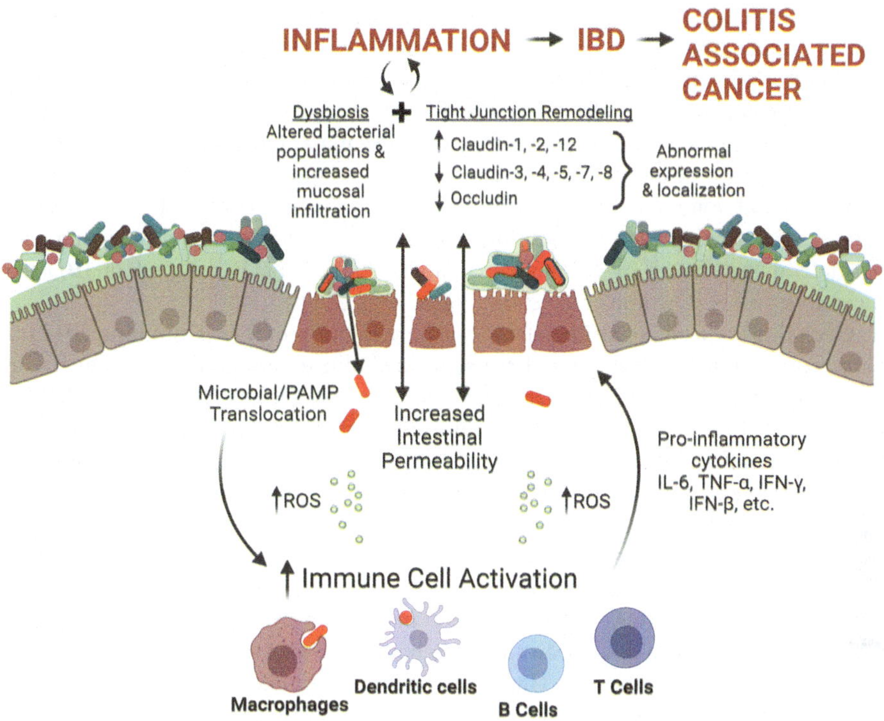

Fig. 1.1 Inflammation in the gut is associated with a dysbiotic microbial environment and abnormal TJ protein expression and localization. Inflammatory conditions promote TJ remodeling and lead to increased intestinal permeability. Translocation of antigenic content across the intestinal barrier activates immune cells, and the release of proinflammatory cytokines further influences TJ dynamics. Chronic inflammation can lead to inflammatory bowel disease (IBD) and increased risk of colitis-associated cancer. Created with BioRender.com

2002; Rosenthal et al. 2010; Tamura et al. 2011). Dysregulation of claudin expression and cellular localization is often seen during intestinal inflammation and other pathologic conditions and may contribute to further gut barrier dysfunction (Lu et al. 2013; Luettig et al. 2015; Ahmad et al. 2017a).

For example, claudin-2 upregulation during intestinal inflammation secondary to infection or autoimmune processes reduces transepithelial electrical resistance (TEER)—a surrogate measurement for gut barrier integrity—and contributes to diarrhea by increasing the flow of water and solutes into the intestinal lumen (Luettig et al. 2015; Tsai et al. 2017; Takehara et al. 2009). This mechanism most likely evolved as a protective response to infection. Interestingly, its overexpression in IBD seems to also have some protective effects against colitis, though the mechanism for this paradoxical claudin-2 regulatory pattern is not fully understood (Dhawan et al. 2011; Ahmad et al. 2014). Claudin-1 is normally expressed at relatively low levels in the colon, but is strongly upregulated in autoinflammatory diseases affecting the digestive tract, such as ulcerative colitis and Crohn's disease (Weber et al. 2008; Pope et al. 2014b). Claudin-3 localizes in the junctional and lateral areas of healthy colonic crypts, and its expression is shown to decrease in IBD (Garcia-Hernandez et al. 2017). Claudin-4 localizes to the lateral and basolateral cell surface and regulates the movement of ions and macromolecules across epithelial surfaces (Deluco et al. 2021). Overexpression of claudin-4 was found to decrease transepithelial conductance, or permeability, by reducing paracellular sodium flow and retention of chloride ions (Van Itallie et al. 2001). Claudin-7 is present along the basolateral membrane of colonocytes, and its expression is reduced in IBD and ulcerative colitis (Oshima et al. 2008; Fujita et al. 2006).

Occludin is essential for the stability and regulation of TJs (Cummins 2012). A close correlation has been found between occludin expression levels and barrier properties in vitro and in vivo (Schulzke et al. 2005). Despite having morphologically intact TJs, occludin knockout mice displayed chronic inflammation and a defective epithelial barrier, indicating that it is more crucial for tight junction stability than assembly (Saitou et al. 2000). Although not essential for the formation of TJs, reduced expression of occludin is associated with increased epithelial permeability and inflammation (Cummins 2012). Junctional adhesion molecule A (JAM-A) is widely expressed in epithelial cells, though its localization is not restricted to TJs. JAM-A is important for maintaining intestinal barrier integrity; JAM-A-deficient mice have been shown to have increased barrier permeability with elevated bacterial translocation (Luissint et al. 2014; Laukoetter et al. 2007).

TJ proteins are also regulated by post-translational modifications, which determine the membrane distribution, localization, and turnover of TJ proteins. Given its critical role in establishing and maintaining TJ integrity, it is unsurprising that occludin is one of the most tightly regulated TJ proteins. Intestinal barrier integrity depends on the phosphorylation of occludin by protein kinase C (PKC) and tyrosine kinase Src, and dephosphorylation by protein phosphatase 2A (PP2A), protein phosphatase 1 (PP1), and protein tyrosine phosphatase 1B (PTP1B) (Jain et al. 2011; Seth et al. 2007; Rao 2009). In homeostatic epithelium, occludin is highly phosphorylated at serine/threonine residues. Tyrosine phosphorylation of occludin at Y398 and Y402 promotes disruption of junctional protein complexes by altering protein–protein interactions with the proteins zonula occludens (ZO)-1, ZO-2, and ZO-3 (Kale et al. 2003; Elias et al. 2009). In the presence of oxidative stress, such as that seen during chronic inflammation, enhanced intestinal permeability is thought to be mediated by increased phosphorylation of occludin tyrosine residues, as well as the redistribution of occludin, ZO-1, and E-cadherin outside of the junctional compartment (Rao et al. 2002). In vitro models of oxidative stress using hydrogen peroxide have also shown that Src family kinases play an important role in promoting tyrosine residue phosphorylation on occludin (Basuroy et al. 2003). Patients with inflammatory bowel syndrome were observed to have delocalized/disassembled occludin that was subsequently degraded by the proteasome; this may occur in a phospholipase D2 (PLD2)-Src dependent manner as seen in the dextran sodium sulfate (DSS) model of colitis (Coëffier et al. 2010; Chelakkot et al. 2017).

Homeostatic regulation of the intestinal epithelial barrier is controlled by complex interactions between TJ proteins and their microenvironment, and dysregulation of these pathways can contribute to chronic inflammation and its sequelae.

1.3 Immune System and Tight Junctions

The innate and adaptive immune systems are intricately involved in the regulation and defense of the intestinal epithelial barrier, working in tandem with TJ proteins to maintain barrier integrity while also clearing infection and cellular debris (Le et al. 2021; Chelakkot et al. 2018). Numerous studies have found that proinflammatory cytokines released by local immune populations during intestinal pathologies, including interleukin 6 (IL-6), interferon γ (IFN-γ), tumor necrosis factor α (TNF-α), and transforming growth factor β (TGF-β), influence the expression and localization of TJ proteins (Capaldo and Nusrat 2009). Chronic inflammatory cytokine exposure has been shown to cause TJ remodeling; however, it is unclear whether these alterations are due to dysfunctional signaling or if they serve as a protective mechanism against further injury (Burstein and Fearon 2008; Fantini and Pallone 2008; Capaldo et al. 2014). During mucosal inflammation, TNF-α promotes the internalization of occludin, resulting in pathogenic gut permeability (Marchiando et al. 2010; Li et al. 2012). IFN-γ has been shown to increase gut permeability by reducing the expression and membrane trafficking of occludin and ZO-1 in an AMP-activated protein kinase A (AMPK)-dependent manner (Utech et al. 2005; Youakim and Ahdieh 1999). IFN-β is implicated in the mechanism by which lipopolysaccharide (LPS)—a bacterial endotoxin and pathogen-associated molecular pattern (PAMP) detected by toll-like receptor 4 (TLR4) in intestinal cells—alters the expression of TJ proteins (Mandel et al. 2012). Whereas LPS does not pass through the gut mucosal and epithelial barrier under physiologic conditions, LPS can permeate through a leaky, dysfunctional barrier in the presence of injury or dysbiosis. In one study, basolateral application of physiologically relevant doses of LPS in the Caco2 colon carcinoma cell line resulted in an increase in TLR4 expression, which in turn increased TJ permeability by activating the TJ adaptor protein focal adhesion kinase (FAK)/MyD88/IL-1R-associated kinase 4 (IRAK-4) signaling pathway (Guo et al. 2015).

Changes in the expression and localization of claudin proteins during inflammation directly affect the permeability and integrity of epithelial barriers. However, the reasons for these changes are not yet fully understood. Claudin expression patterns are affected by inflammatory stimuli and are associated with a state of gut barrier dysregulation, though whether claudins exacerbate or protect against inflammation is unclear given the complexity of their regulation and involvement in differential cell signaling pathways and the context dependence of these mechanisms. Because claudins are involved in a multitude of signaling pathways in which their involvement—like their expression and localization—differs across cell types, it is difficult to determine whether these changes are inherently protective or pathologic, or simply consequences of signaling pathways affected by the inflammatory insult.

For instance, it was previously believed that claudin-2 upregulation in IBD contributed to IBD pathogenesis, but claudin-2 is now known to be essential to gut barrier homeostasis and can protect against inflammation (Ahmad et al. 2014). Given their regulation by inflammatory cytokines and opposing characteristics in the TJ complex, claudin-2 and claudin-4 have been of particular interest in IBD research (Prasad et al. 2005). During mucosal inflammation, claudin-4 expression decreases while claudin-2 expression increases, resulting in increased gut permeability and greater susceptibility to further inflammatory insults (Oshima et al. 2008). These alterations in protein expression are thought to be the direct result of proinflammatory cytokine signaling; Capaldo et al. have demonstrated that intestinal epithelial cell lines treated with IFN-γ and TNF-α have reduced assembly of claudin-4 (Capaldo et al. 2014). Given its importance in mitigating colitis by interacting with the immune microenvironment of the inflamed colon, claudin-2 may be a promising therapeutic target for minimizing the extent of injury during an IBD flare. In mouse models of chemically induced colitis, overexpression of claudin-2 can protect mice from severe inflammatory injury by increasing the local population of immunosuppressive regulatory T cells (Tregs) and anergic macrophages, as well as promoting TGF-β production and secretion by colonocytes (Ahmad et al. 2014).

1.4 Gut Dysbiosis

Gut dysbiosis is an alteration in the composition of gut microbiota often present in gastrointestinal diseases, including colitis and colon cancer (Schaubeck et al. 2016; Nakatsu et al. 2015; Henao-Mejia et al. 2012; Zeng et al. 2017); IBD and colorectal cancer (CRC) patients have been found to have dysbiosis and a dysfunctional gut barrier with epithelial hyperpermeability and increased mucosal bacteria (Yu 2018).

Under physiologic conditions, constant but subtle interactions occur between the luminal contents of the gastrointestinal tract and the intestinal epithelium. Pattern-recognition receptors (PRRs) such as Toll-like receptors (TLRs) are widely expressed on both epithelial cells and immune cells. PRRs and TLRs facilitate the recognition of viral and bacterial particles, also known as pathogen-associated molecular patterns (PAMPs), or damage-associated molecular patterns (DAMPs) derived from injured or dying cells (Kawasaki and Kawai 2014). PRRs can also differentiate between commensal and pathogenic microbes. Intestinal mucus, composed of heavily O-glycosylated mucin proteins secreted by specialized mucus-producing intestinal cells known as goblet cells, prevents microbial translocation by trapping microbes and water-insoluble antigens to limit their contact with the underlying epithelium and immune cells (Cornick et al. 2015). Any changes in the expression of TJ proteins can alter the function of TJ complexes and therefore affect gastrointestinal barrier functionality, potentially increasing the risk of gut microbiota translocation through the intestinal barrier into the systemic circulation. Once disruption occurs, luminal particles can rapidly pass into the stroma, initializing the inflammatory cascade in the gut. Activation of PRRs by PAMPs or DAMPs triggers downstream signaling pathways that promote inflammatory responses and immune cell recruitment. Therefore, any disruption in the homeostasis of commensal bacterial populations can trigger swift immune responses to prevent or respond to translocation of bacteria or their products across a damaged or disrupted intestinal barrier.

1.5 Inflammatory Bowel Disease

Inflammatory bowel disease (IBD) is a group of disorders including Crohn's disease and ulcerative colitis that cause mucosal inflammation and injury throughout the GI tract, causing disruption and instability of the gut barrier that can lead to dysbiosis, local infection or sepsis, and inflammatory cancers (Yu 2018; Genua et al. 2021). Although the pathogenesis of IBD is not completely understood, multiple studies have demonstrated that genetic factors, high-fat diet, past infections or other perturbations in gut microbiota, and various environmental factors can influence the development of IBD (Loddo and Romano 2015; Paik et al. 2013; Gruber et al. 2015; Ananthakrishnan et al. 2018). Despite ongoing research, it remains unclear whether the changes observed in TJs in IBD are causal, leading to abnormal epithelial barrier integrity and aberrant immune responses, or a result of inflammation that leads to TJ disruption.

1.5.1 Crohn's Disease

Crohn's disease (CD) is a type of Th1-mediated colitis that causes transmural inflammation throughout the GI tract, but primarily in the terminal ileum and colon. CD patients commonly present with abdominal pain and diarrhea (with or without hematochezia). Severe cases can lead to weight loss, anemia, and fatigue due to malabsorption resulting from extensive intestinal damage (Peppercorn 2022). Poorly controlled, chronic CD may lead to the formation of esophageal or intestinal strictures and/or fistula formation. Strictures are areas of fibrosis and narrowing of the GI tract lumen that can cause obstruction, infection, and perforation (Peppercorn 2022). Fistulas are pathological openings or passages caused by extensive transmural ulceration that permit the passage of GI luminal contents into the bladder (enterovesical fistula), perianal region, abdominal cavity, or systemic circulation. Extraintestinal manifestations of CD include, but are not limited to, hepatobiliary disease, uveitis, arthritis, and kidney stones (Peppercorn 2022).

Alterations in the expression of several TJ proteins have been observed in patients with CD (Cuzic et al. 2021). In intestinal biopsies from patients with active CD, claudin-3, -5, -8, and occludin were found to be decreased, while claudin-2 and -12 were found to be moderately increased, especially in the ileum (Garcia-Hernandez et al. 2017). Further, claudin-2 was found to be strongly expressed in the ileal region of approximately 50% of patients with active CD; of note, approximately 80% of CD patients tend to present with distal ileal involvement (Zeissig et al. 2007). Claudin-2 expression in CD patients was found to be associated with disrupted TJs (Zeissig et al. 2007). Although the overall levels of tricellulin—a TJ protein found at the intersection of epithelial or endothelial cells that is essential for maintaining an effective barrier against macromolecules (Krug et al. 2009)—remained unchanged in patient samples, its localization was found to change from the basal region of the crypt to the apical surface (Krug et al. 2018). One study noted that angulin-1, a regulator of tricellulin localization, was downregulated in CD, demonstrating a possible mechanism for tricellulin delocalization and therefore disrupted barrier integrity in CD (Hu et al. 2020).

1.5.2 Ulcerative Colitis

Ulcerative colitis (UC) is a form of IBD that causes Th2-mediated mucosal inflammation localized to the large intestine (Xu et al. 2014). UC patients often present with gradual-onset abdominal pain, bloody diarrhea (unlike CD in which diarrhea may not contain blood), incontinence, or constipation. Like CD, UC may also be accompanied by fatigue and weight loss secondary to anemia (Ungaro et al. 2017). Iron-deficiency anemia is of particular importance in UC, as UC can cause bleeding as a result of autoimmune-mediated colon epithelial ulceration and erosion. Though not as common as in CD, patients may experience colon perforation secondary to toxic megacolon (Autenrieth and Baumgart 2012). Toxic megacolon is the result of severe inflammation extending into the muscularis mucosa of the colon, typically in the ileocecal region, causing increased colon (≥6 cm) or cecal (>9 cm) diameter (Autenrieth and Baumgart 2012). UC may also be accompanied by extraintestinal manifestations similar to those seen in CD; however, patients with UC are more commonly affected by primary sclerosing cholangitis than patients with CD (Tanaka and Mertens 2016).

Several studies have identified TJ proteins as important players in the pathogenesis of UC. In the sigmoid colon of early UC patients, though the epithelium looked intact, there was higher conductance of solutes across the epithelium, which correlated with a higher degree of inflammation (Paine 2014). Epithelial resistance and TJ depth in inflamed samples were reduced by 80% (Paine 2014). As in CD, claudin-2 was found to be more highly expressed in colon samples from patients with UC compared to normal controls, and claudin-2 expression levels were also found to correlate with disease severity (Luettig et al. 2015). On the other hand, claudin-3, -4, and -7, as well as occludin, appear to be downregulated on the apical surface of the colon epithelium, and there is evidence that claudin-4 may localize outside of the TJ in UC patients (Oshima et al. 2008). A strong correlation is present between disease severity and changes in TJ structure. In sigmoid colon samples from patients with active UC, upregulation of claudin-12—a "tightening" claudin—may be associated with the degree of inflammation (Landy et al. 2016). Similarly, a primary defect in barrier function was postulated to explain the elevated expression of claudin-18 in UC patients compared to controls (Zwiers et al. 2008). In a study of two different mouse models of colitis resembling human UC, occludin knockout resulted in reduced susceptibility to inflammatory injury from colitis due to its pro-apoptotic functions (Kuo et al. 2019). The reduction of occludin expression in response to inflammation inhibits epithelial apoptosis by reducing caspase-3 expression (Kuo et al. 2019). Moreover, it has been shown that claudin-1-overexpressing mice were more susceptible to experimental DSS-induced colitis as a result of impaired goblet cell differentiation, delayed intestinal epithelial recovery, sustained inflammation, and crypt hyperplasia, possibly resulting from the activation of Notch signaling by matrix metalloproteinases-9 (MMP-9)/extracellular signal-regulated kinase (ERK) (Pope et al. 2014b; Gowrikumar et al. 2019).

1.6 Barrier Dysfunction and Tight Junction Regulation in Gastrointestinal Cancers

Regulation of TJ protein expression is crucial for maintaining proper physiological function and permeability of the intestinal epithelium. In addition to their primary role of regulating the paracellular transport of ions and small molecules in endothelial and epithelial tissues, TJ proteins are also known to interact with signaling pathways involved in carcinogenesis, including those affecting cell proliferation, malignant transformation, and metastasis. It has long been the accepted paradigm that the loss of TJ protein expression and the ensuing impairments in cell-cell adhesion allow tumor cells to detach from primary tumors, paving the way for local invasion and subsequent distant metastasis. In addition, the loss of TJ adhesion disrupts cellular polarity and promotes dedifferentiation, another characteristic of cancer cells (Martin and Jiang 2009). A number of studies have found that depletion of ZO-1, occludin, claudin-7, and claudin-16 may hasten tumor growth (Martin et al. 2010; Kominsky et al. 2003).

More recently, however, this paradigm has been called into question by research demonstrating that overexpression of some TJ proteins, such as JAM-A, claudin-1, and claudin-7, is correlated with tumor progression and metastasis. Though it may seem contradictory, these data may point to novel functions of TJ proteins. In other words, it is the *abnormal* expression of TJ proteins that characterizes tumor progression rather than a simple loss of overall expression. More studies are needed to further elucidate the signaling pathways and regulatory mechanisms involving TJ proteins in tumorigenesis and cancer progression and how they may differ across cancer and tissue types. Table 1.1 highlights several findings that have identified roles for TJ overexpression/loss in tumorigenesis and other diseases.

1.7 Tight Junction Proteins in Cancer Signaling

1.7.1 Protein Kinase C (PKC)

As previously noted, TJ regulation in response to inflammatory cytokine stimulation is in part mediated by protein kinase C (PKC), a member of the serine-threonine kinase family (Nishizuka 1992). PKC influences the assembly and disassembly of TJ

Table 1.1 Role of tight junction (TJ) proteins in various diseases and the underlying mechanisms. Abbreviations: JAM - junctional adhesion molecule, PI3K - phosphoinositide 3-kinase, MAPK - mitogen-activated protein kinase, TGF-β1 - transforming growth factor β1, ROCK - Rho-associated protein kinase, LRP5 - low-density lipoprotein receptor-related protein 5, AKT - protein kinase B (PKB), CCND1 - cyclin-D1, JAML - junctional adhesion molecule-like, NF-κB - nuclear factor kappa-light-chain-enhancer of activated B cells, ERK - extracellular signal-related kinase, Wnt - Wingless-related integration site, EGFR - epidermal growth factor receptor, STAT3 - signal transducer and activator of transcription 3, AMPK - adenosine monophosphate-activated protein kinase, NOS - nitric oxide synthase, NO - nitric oxide, ZONAB - zonula occludens-1 (ZO-1)-associated nucleic acid binding protein, MMP - matrix metalloprotease, FAK - focal adhesion kinase

S. No	Tight Junction Proteins	Pathogenesis	Expression	Mechanism	References
1.	JAM-A	**Cancer:**		PI3K/MAPK signaling, Notch signaling, TGF-β1 signaling, ROCK pathways	Rosager et al. (2017), McSherry et al. (2009), Naik et al. (2008), Brennan et al. (2013), McSherry et al. (2011), Götte et al. (2010), Goetsch et al. (2013), Ikeo et al. (2015), Zhang et al. (2013), Zhao et al. (2014), Lampis et al. (2021)
		Brain	↑		
		Breast	↑↓		
		Gastric	↑		
		Lung	↑		
		Endometrial	↓		
		Pancreatic	↓		
		Colorectal cancer	↓		
		Hereditary diseases:			Coyne et al. (2002)
		Cystic fibrosis	↓		
		Viral infection:			Forrest et al. (2003)
		Reoviral infection (hydrocephalus, encephalitis)	↑		
2.	JAM-B	**Cancer:**			
		Pancreatic	↑	c-Src signaling	Zhang et al. (2020)
		Glioma	↑	Notch pathway	Qi et al. (2013, 2014)
3.	JAM-C	**Cancer:**		LRP5/AKT/β-catenin/CCND1 signaling, PI3K/MAPK signaling	Fuse et al. (2007), Hao et al. (2014), Arcangeli et al. (2012), Ghislin et al. (2011), Langer et al. (2011), Leinster et al. (2013), Zhang et al. (2018), Ebnet (2017), Sheng et al. (2020)
		Fibrosarcoma	↑		
		Lung	↑		
		Melanoma	↑		
		Ovarian	↑		

		Inflammation:			Aurrand-Lions et al (2005)
		Pulmonary inflammation	↑		
4.	JAML	**Cancer:**		P38 signaling	Fang et al. (2021)
		Gastric	↑		
3.	Coxsackievirus and adenovirus receptor (CAR)	**Cancer:**		MyD88/IRAK-4/NF-κB, ERK1/2 pathway, estrogen signaling	Martin et al. (2005), Brüning et al. (2005), Giaginis et al. (2008), Chen et al. (2013), Saito et al. (2014), Reimer et al. (2007), Giaginis et al. (2010), Vindrieux et al. (2011), Maekawa et al. (2007)
		Breast	↑		
		Endometrial	↑		
		Lung	↑		
		Oral	↑		
		Ovarian	↑		
		Thyroid	↑		
		Inflammation:			
		Autoimmune myocarditis	↑		Ito et al. (2000)
		Cystic fibrosis	↑		Sharma et al. (2017)
		Acute and chronic lung inflammation	Phosphorylated		Morton et al. (2016)
4.	Claudin-1	**Cancer:**		Notch signaling, Wnt signaling, PI3K signaling	Leech et al. (2015), Oliveira and Morgado-Díaz (2007), Itoh and Bissell (2003), Akasaka et al. (2010), Liu et al. (2012), Achari (2015), Lee et al. (2005), Kinugasa et al. (2007), Caruso et al. (2014), Bhat et al. (2012), Dhawan et al. (2005), Huo et al. (2009), Resnick et al. (2005), Singh et al. (2012), Sheehan et al. (2007), Tzelepi et al. (2008), Lv et al. (2017), Miwa et al. (2001), González-Mariscal et al. (2008)
		Breast	↑↓		
		Cervical	↑	P38, c-Abl-ERK MAPK signaling	
		Colorectal	↑↓		
		Gastric	↑		
		Liver	↑		
		Oral	↑		
		Ovarian	↑		
		Prostate	–		

(continued)

Table 1.1 (continued)

S. No	Tight Junction Proteins	Pathogenesis	Expression	Mechanism	References
		Thyroid	↑		
		Inflammation:			Tokumasu et al. (2016)
		Inflammatory bowel disease (IBD)	↑		
		Atopic dermatitis (AD)	↓		
		Hereditary diseases:			Coyne et al. (2002)
		Cystic fibrosis	–		
5.	Claudin-2	**Cancer:**			
		Breast	↑↓	EGFR/MEK/ERK pathway	González-Mariscal et al. (2008), Tabariès et al. (2012), Kimbung et al. (2014), Tabariès et al. (2012, 2015), Buchert et al. (2010), Kinugasa et al. (2007), Ikari et al. (2012, 2014), Hintsala et al. (2013)
		Colorectal	↑	PI3K signaling	
		Lung	↑		
		Skin	↑		
		Prostate	↓		
		Inflammation:			
		Crohn's disease	↑		Zeissig et al. (2007)
		Collagenous colitis	↑		Bürgel et al. (2002)
6.	Claudin-3	**Cancer:**		EGFR/MEK/ERK pathway	Hewitt et al. (2006), Kolokytha et al. (2014), de Souza et al. (2013), Konecny et al. (2008), Santin et al. (2007), Satake et al. (2008), Choi et al. (2007), Agarwal et al. (2005), Sheehan et al. (2007)
		Breast	↑	PI3K/AKT signaling	
		Colorectal	↑	Wnt signaling	
		Endometrial	↑↓	STAT3	
		Gastric	↑		
		Kidney	↑		
		Lung	↑		
		Ovarian	↑		

		Prostate	–		
		Uterine	↑		
		Inflammation:			Zeissig et al. (2007), Luettig et al. (2015)
		Crohn's disease	↓		
		Bacterial toxins:			McClane (2001)
		Clostridium perfringens enterotoxin-mediated diarrhea	↓		
7.	Claudin-4	**Cancer:**		ERK pathway	
		Breast	↑	AMPK pathway	Hewitt et al. (2006), Konecny et al. (2008), Suren et al. (2015), Pan et al. (2007), Singh et al. (2010), Hwang et al. (2014), Agarwal et al. (2005), Sheehan et al. (2007)
		Endometrial	↑		
		Gastric	↑↓		
		Kidney	↑		
		Lung	↑		
		Nasopharyngeal	↑		
		Ovarian	↑		
		Pancreatic	↑		
		Uterine	↑		
		Inflammation:			Bürgel et al. (2002)
		Collagenous colitis	↓		
		Hereditary diseases:			Coyne et al. (2002)
		Cystic fibrosis	↓		
		Bacterial toxins:			McClane (2001)
		Clostridium perfringens enterotoxin-mediated diarrhea			

(continued)

Table 1.1 (continued)

S. No	Tight Junction Proteins	Pathogenesis	Expression	Mechanism	References
8.	Claudin-5	**Cancer:**	↓	NOS/NO/ZONAB, leading to enhanced permeability	Miao et al. (2015), Ma et al. (2014), Liu et al. (2015)
		Glioma			
9.	Claudin-6	**Cancer:**	↑	Overexpression leads to MMP-2 activation	Torres-Martínez et al. (2017)
		Gastric			
10.	Claudin-7	**Cancer:**		ERK/MAPK signaling pathway	Singh et al. (2010), Lee et al. (2005), Philip et al. (2015), Hewitt et al. (2006), Suren et al. (2015), Sheehan et al. (2007), Tzelepi et al. (2008)
		Breast	↓	Wnt signaling	
		Cervical	↑	Integrin/FAK signaling pathway	
		Colon	↑		
		Gastric	↑		
		Liver	↑		
		Lung	↑		
		Nasopharyngeal	↑		
		Ovarian	↑		
		Pancreatic	↑		
		Prostate	↑		
		Thyroid	↑		
		Inflammation:			Luettig et al. (2015)
		Crohn's disease	–		
		Ulcerative colitis	↓		
		Celiac disease	–		
11.	Claudin-8	**Cancer:**		–	
		Colorectal	↓		Grone et al. (2007)
		Renal oncocytoma	↓		Kim et al. (2009), Osunkoya et al. (2009)
12.	Claudin-9	**Cancer:**		Associated with overexpressed MMP-12, weak blood vascular	
		Lung	↑		Sharma et al. (2016)

		Pituitary oncocytoma	↑	endothelium, actin cytoskeleton reorganization, paracellular permeability	Hong et al. (2014)
13.	Claudin-10	**Cancer:**			
		Lung	↑	c-fos pathway	Zhang et al. (2013)
		Cholangiocarcinoma	↓	–	Németh et al. (2009)
14.	Claudin-11	**Cancer:**		TGF-β, ERK, p38	Lin et al. (2013)
		Gastric	↑		
15.	Claudin-12	**Cancer:**		–	Grone et al. (2007)
		Colorectal	↑		
16.	Claudin-14	**Cancer:**			Li et al. (2016)
		Hepatocellular	↓		
17.	Claudin-15	**Cancer:**			
		Malignant pleural mesothelioma	↓	Overexpressed Claudin-15 serves as potential antiproliferative function	Chaouche-Mazouni et al. (2013)
		Colitis associated cancer	↑	–	Arimura et al. (2011)
		Colon	↑	Associated with MMP-2 and -9 activation	Takehara et al. (2009)
18.	Claudin-16	**Cancer:**			Kuo et al. (2010), Rangel et al. (2003), Men et al. (2015)
		Breast	↑		
		Ovarian	↑		
		Renal	↑	–	
		Hereditary diseases:			Simon et al. (1999)
		Familial hypomagnesemia	Mutation		
19.	Claudin-17	**Cancer:**		–	Gao et al. (2013)
		Gastric	↓		

(continued)

Table 1.1 (continued)

S. No	Tight Junction Proteins	Pathogenesis	Expression	Mechanism	References
20.	Claudin-18	**Cancer:**		p-AKT	Akizuki et al. (2017)
		Lung	↓		
21.	Claudin-20	**Cancer:**		–	Martin et al. (2013)
		Breast	↑		
22.	Occludin	**Cancer:**		PI3K signaling	González-Mariscal et al. (2008, Tzelepi et al. (2008)
		Thyroid	Diverse expression	MAPK signaling	
		Inflammation:			
		Crohn's disease	↓		Bürgel et al. (2002)
		Ulcerative colitis	↓		Förster et al. (2007)
		Celiac disease	↓		Luettig et al. (2015)
		Hereditary diseases:			Coyne et al. (2002)
		Cystic fibrosis	↑		
		Vision loss:			Felinski and Antonetti (2005)
		Diabetic retinopathy	↑		
23.	ZO-1	**Cancer:**		PI3K signaling	González-Mariscal et al. (2008, Itoh and Bissell (2003)
		Breast	↓	MAPK signaling	
		Inflammation:			Zeissig et al. (2007)
		Crohn's disease	↓		
		Vision loss:			Felinski and Antonetti (2005)
		Diabetic retinopathy	↑		

proteins, most notably occludin (Andreeva et al. 2001). Individuals with pancreatic cancer show increased expression levels of PKCα, PKCβ1, PKCδ, and PKCι in cancer tissue compared to normal adjacent tissue (El-Rayes et al. 2008). In pancreatic cancer, PKC signaling has been shown to regulate claudin-1, -4, -7, and -18, occludin, JAM-A, and ZO-1 and -2 (Kyuno et al. 2013; Tanaka et al. 2011; Yamaguchi et al. 2010). PKCs are also found to be mutated in gastric cancer, with loss of function (LOF) mutations being the most commonly observed (Yamaguchi et al. 2010). LOF mutations in the catalytic or kinase domains of PKC result in impaired kinase activation and function (Antal et al. 2015). LOF mutations in PKC have also been linked to the failure of ZO-1 to translocate to the TJ region, leading to problems in TJ integrity and an increased risk of tumor invasion (Cario et al. 2004). In hepatocellular carcinoma, activation of c-Abl tyrosine kinase-PKCδ signaling downstream of increased claudin-1 expression increases tumor invasiveness (Lee et al. 2015) (Fig. 1.2).

1.7.2 Anoikis Resistance Pathways

Resistance to anoikis, apoptotic cell death induced by the separation of cells from the extracellular matrix (ECM), is critical for the anchorage-independent growth of cancer cells leading to invasion and metastasis. It often involves changes in cell-cell interactions, interactions with the ECM, and aberrant activation of pro-survival pathways. ECM adhesion itself activates numerous pro-survival pathways, including phosphoinositide 3-kinase (PI3K)/AKT, ERK, and growth factor signaling (Paoli et al. 2013). Activation of these pathways in the tumor microenvironment irrespective of ECM adhesion results in cell survival despite anchorage-independent growth. For instance, in colon adenocarcinoma cell lines, stimulation of the epidermal growth factor receptor (EGFR) by epidermal growth factor (EGF) released by cancer-associated fibroblasts increases the expression of claudin-2, which then promotes anchorage-independent tumor growth via mitogen-activated protein kinase (MAPK) and PI3K/AKT signaling pathways (Dhawan et al. 2011). Similarly, Jose et al. found that activation of EGF signaling in HT-29 colon adenocarcinoma cells accelerated cell motility and anchorage-independent growth by increasing claudin-3 expression downstream of ERK1/2 and PI3K-AKT signaling (de Souza et al. 2013). Singh et al. found that claudin-1 overexpression in colon cancer cells is linked to anoikis resistance, mediated through direct interactions between claudin-1 and Src leading to the activation of anti-apoptotic PI3K/AKT/Beclin-2 (Bcl2) signaling (Singh et al. 2012). Although various claudins and JAMs have been found to contribute to cancer invasiveness by increasing anoikis resistance, the same cannot be said for occludin. Studies of breast cancer cell lines have shown that occludin overexpression increases sensitivity to apoptosis and increased the incidence of anoikis (Osanai et al. 2007).

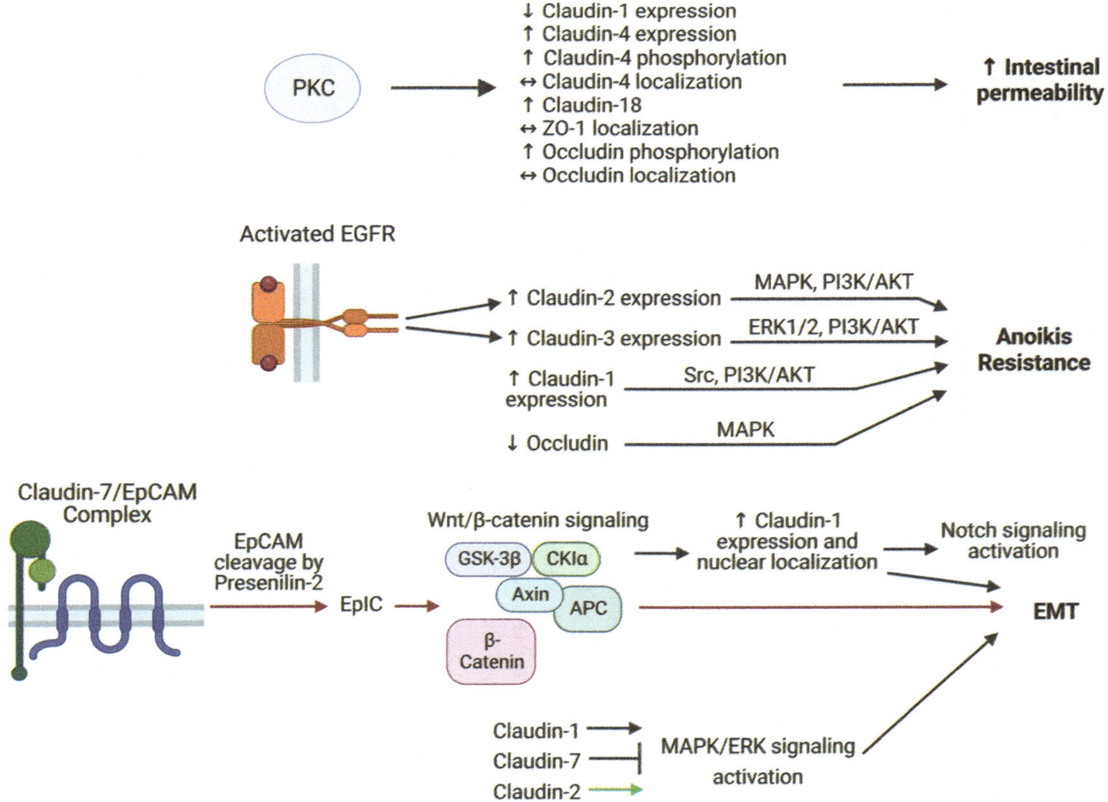

Fig. 1.2 Summary of several oncogenic signaling cascades that involve or regulate TJ proteins in gastrointestinal cancers to increase intestinal epithelial permeability, anoikis resistance, and epithelial-to-mesenchymal transition (EMT): features that promote tumor survival and invasiveness. Abbreviations: PKC - protein kinase C, ZO-1 - zonula occludens 1, EGFR - epidermal growth factor receptor, MAPK - mitogen-activated protein kinase, PI3K - phosphoinositide-3 kinase, EpCAM - epithelial cellular adhesion molecule, EpIC - EpCAM intracellular domain (cleaved EpCAM), GSK3β - glycogen synthase kinase-3 β, CKIα - casein kinase 1 α, APC - adenomatous polyposis coli, EMT - epithelial-to-mesenchymal transition. Created with BioRender.com

1.7.3 Notch and Wnt Signaling

Notch and Wingless-related integration site (Wnt)/β-catenin signaling are involved in the regulation of intestinal epithelial cell renewal and differentiation (Fevr et al. 2007). Hyperactivation of the Wnt/β-catenin signaling pathway in numerous cancers results in elevated expression of genes involved in cell survival, proliferation, and invasion, including c-myc, cyclin-D1, MMP-7, T-cell factor 1 (Tcf1), and ephrin type-B receptor 2 (EphB2) (van de Wetering et al. 2002; Rodilla et al. 2009). Over 70% of colorectal cancers have mutations in the adenomatous polyposis coli (APC) gene, resulting in Wnt signaling activation. Dhawan et al. have shown that caudal-related homeobox (Cdx) transcription factors in conjunction with β-catenin and Tcf downstream of Wnt signaling induce claudin-1 transcription in human colon cancer cells (Bhat et al. 2012). The Notch pathway also interacts with the Wnt/β-catenin and PI3K/AKT/mammalian target of rapamycin (mTOR) pathways to promote EMT (Sharma et al. 2022). Claudin-1 has also been shown to regulate Notch signaling. Overexpression of claudin-1 in a goblet cell-like cell line (LS174T) resulted in an increase in expression of the Notch intracellular domain (NICD), which was mediated by MMP-9 and ERK signaling (Pope et al. 2014b). In the same study, similar effects on Notch signaling, MMP-9, and ERK activation were also seen in a claudin-1-overexpressing transgenic mouse model (Rosenthal et al. 2010). Claudin-7 has been shown to interact with epithelial cellular adhesion molecule (EpCAM) to promote its cleavage, after which the resulting EpCAM intracellular domain (EpIC) interacts with β-catenin and other factors to interfere with E-cadherin-mediated cell adhesion or to transcribe various stemness-related genes and promote EMT (Philip et al. 2015). Nuclear localization of claudin-1 and β-catenin was detected in liver metastatic lesion samples indicating that claudin-1 may support the translocation of membranous β-catenin to activate its target genes, resulting in robust growth and/or survival of malignant cells (Dhawan et al. 2005).

1.7.4 MAPK/ERK Signaling

The mitogen-activated protein kinase (MAPK) cascade is a major signaling pathway whose downstream effects include regulation of cell proliferation, survival, apoptosis, motility, transcription, metabolism, and differentiation. In several studies, the MAPK/ERK signaling pathway was shown to impact the expression and distribution of various TJ proteins, including ZO-1, ZO-2, ZO-3, claudin-1, claudin-2, and claudin-4 (Murata et al. 2005; Lipschutz et al. 2005). One study found that calcium oxalate crystals, components of kidney stones, activate p38 in renal tubular epithelial cells, resulting in downregulation of ZO-1 and occludin and leading to TJ disruption and renal tubular dysfunction (Peerapen and Thongboonkerd 2013). ERK inhibition in Caco-2 cells grown in a monolayer reduces the localization of occludin and ZO-1 at TJs by regulating the threonine residue phosphorylation of occludin by PKCζ and PP2A (Dhawan et al. 2011; Ahmad et al. 2014; Aggarwal et al. 2011). Further, claudin-7 has been noted to suppress ERK/MAPK signaling in non-small cell lung cancer models, leading to reduced cell mobility and invasion (Xia et al. 2014). Similarly, overexpression of claudin-7 in colon cancer cells decreases proliferation and invasion by inhibiting ERK and Src signaling (Bhat et al. 2015). In hepatocellular carcinoma, claudin-1 can promote EMT by activating the c-Abl-Ras-Raf-1/ERK1/2 signaling pathway (Suh et al. 2013). Claudin-2 overexpression in lung cancer has also been shown to activate the EGFR/MEK/ERK signaling pathway. c-Fos, a downstream target of EGFR/MEK/ERK, upregulates transcription of claudin-2 by binding to the activator protein 1 (AP-1) promoter site (Ikari et al. 2012).

1.8 Tight Junctions in Various Sporadic GI Cancers

1.8.1 Esophageal Adenocarcinoma

Claudin expression has been explored as a biomarker for progression from Barrett's esophagus—pre-malignant dysplasia of the esophageal epithelium due to chronic gastroesophageal reflux disease (GERD) and other risk factors—to esophageal adenocarcinoma. Claudin-2, -3, -4, and -7 are upregulated in esophageal adenocarcinoma compared to normal esophageal epithelium (Takala et al. 2007), and claudin-18 has been detected in Barrett's esophagus (Fujita et al. 2006). Oshima et al. showed that in normal human esophageal stratified squamous epithelium, claudin-1, claudin-4, occludin, and ZO-1 co-localize in TJs, though their expression patterns vary along the apicobasal axis (Oshima et al. 2012). There is little to no expression of ZO-1 in the basal or intermediate layers, while claudin-1 is detectable primarily in the suprabasal and intermediate layers (Fujita et al. 2006). Exposure of normal human bronchial epithelial cell-derived stratified squamous epithelium to acid (simulating acute acid reflux into the esophagus) caused a reduction in claudin-4 expression, co-localization with ZO-1, and localization in the TJ (Fujita et al. 2006). These findings suggest that even acute tissue damage with acid causes alterations in TJ proteins that result in barrier dysfunction and affect the ability of esophageal epithelium to withstand future insults. Indeed, claudin-1 mRNA was elevated in the esophageal mucosa of patients with erosive reflux disease, and elevated mRNA levels of claudin-1 and claudin-2 were associated with basal cell hyperplasia in GERD patients, though it is possible that elevation of these proteins may serve more as a biomarker of active or worsening disease rather than as a driver of pathogenesis (Mönkemüller et al. 2012).

During the early stage of normal epithelial apicobasal polarization, ZO-1 co-localizes with E-cadherin and binds with α- and β-catenin (Ando-Akatsuka et al. 1999). However, in esophageal adenocarcinoma, occludin expression is greatly decreased and ZO-1 localization shifts to basal surfaces compared to normal adjacent tissue (Oshima et al. 2012). According to more recent studies, low expression of occludin was associated with a higher histologic grade and poorer prognosis in patients with esophageal squamous cell carcinoma (Qin et al. 2017). Given that these proteins have been linked to tumor differentiation, it is possible that occludin and

ZO-1 may be key players in the cascade of epithelial barrier dysfunction and eventually dysplastic progression in the esophageal epithelium (Kimura et al. 1997).

1.8.2 Pancreatic Cancer

Normal pancreatic ductal and acinar structures express claudin-1, claudin-2, claudin-3, claudin-4, and claudin-7, while endocrine cells in the islets of Langerhans only express claudin-3 and claudin-7 (Yamaguchi et al. 2010; Borka et al. 2007). Disruption of TJs in the pancreatic duct is observed in different clinical states including pancreatic ductal adenocarcinoma (PDAC), pancreatic neuroendocrine tumors (PanNET), and pancreatitis. Claudin-4 and claudin-18 are the most abundantly expressed claudins in pancreatic tumors, with overexpression occurring even in precancerous pancreatic intraepithelial neoplasia (PanIN), intraductal papillary neoplasia (IPMN), and mucinous cystic neoplasia (MCN) (Tanaka et al. 2011; Kojima et al. 2012). Interestingly, claudin-4 overexpression in pancreatic cancer tissues and cell lines has been shown to decrease invasiveness and metastatic potential (Michl et al. 2003). In PANC-1 cells, TGF-β was able to downregulate claudin-4. Claudin-4 expression was also lowered by dominant-negative Ras expression and inhibition of both MAPK/ERK and PI3K signaling pathways (Michl et al. 2003). Despite also being highly expressed in PDAC, claudin-18 expression is more prominent in well-differentiated pancreatic neoplasms and is associated with good survival in PDAC patients (Kopp et al. 2018).

Alterations in claudin-7 expression and localization are seen in both PDAC and PanNET. In poorly differentiated human PDAC tissues, claudin-7 immunoreactivity in TJ regions was found to be lower than in well-differentiated tumors (Aiello et al. 2018). Moreover, claudin-7 localizes to glycolipid-enriched membrane microdomains rather than the TJ fraction, and it activates the PI3K/AKT pathway, upregulates mesenchymal genes, and promotes the translocation of EMT transcription factors into the nucleus (Thuma et al. 2016). Interestingly, microRNA (miRNA)-mediated suppression of claudin-7 in PDAC-derived exosomes inhibits the invasiveness and motility of pancreatic cancer cells (Kyuno et al. 2019, 2021). In general, claudin-7 overexpression is seen in populations of pancreatic tumors that are highly proliferative, likely due to its regulation of EpCAM-mediated tumorigenic activities, including proliferation, migration, and resistance to apoptosis (Nubel et al. 2009).

In PDAC, claudin-1 expression is TNF-α-dependent, and it has been hypothesized that claudin-1 links cell adhesion and TNF-α-dependent growth signals in pancreatic cancer cells. As pancreatic tumors progress, claudin-1 localization at TJs decreases (Kondo et al. 2008). TGF-β1 present in the pancreatic tumor microenvironment triggers EMT (Principe et al. 2021). PKC signaling controls claudin-1 during EMT in pancreatic cancer through the EMT transcriptional repressor, Snail, and MAPK/ERK-dependent mechanisms (Kyuno et al. 2013).

Tricellulin is a component of tricellular TJs that is expressed in normal and cancerous pancreatic exocrine tissues, but not endocrine cells. In PDAC, tricellulin expression is elevated in more differentiated (lower grade) tumors (Korompay et al. 2012). Tricellulin expression is controlled by c-Jun N-terminal kinase (JNK) (Kojima et al. 2013). On the other hand, the bicellular TJ protein marvelD3, which is found in normal and cancerous pancreatic ductal epithelium, is transcriptionally downregulated in poorly differentiated pancreatic cancer cells. Similarly to claudins, marvelD3 is downregulated during Snail-induced EMT in PDAC cells (Kojima et al. 2011). ZO-1 expression is elevated in PDAC cells, but its cellular localization may vary depending on the tumor microenvironment and cancer cell differentiation. In PDAC tumors with higher invasive potential, the disassociation of ZO-1 from apicolateral TJs and its translocation to the nucleus and cytoplasm occurs downstream of EGFR activation (Takai et al. 2005).

1.8.3 Gastric Adenocarcinoma

In gastric adenocarcinoma, there is a decrease in claudin-1, -3, -5, and -7 (Wang and Yang 2015; Soini et al. 2006; Jun et al. 2014). In general, elevated claudin-2-expression is a risk factor for gastric cancer development (Lin et al. 2013). Increased expression of claudin-4 is a biomarker for poor gastric cancer prognosis; one study has shown that upregulated claudin-4 in gastric adenocarcinoma also activated MMP-2 and MMP-9 expression, both of which promote invasion and metastasis (Hwang et al. 2014). Claudin-6 is also elevated in gastric cancer and has been shown to influence the yes-associated protein 1 (YAP1)-Snail1 axis to promote gastric cancer proliferation and EMT (Yu et al. 2019). Reduction of claudin-11 expression by elevated miR-421 was found to enhance gastric cancer cell proliferation, invasion, and metastasis (Li et al. 2012; Yang et al. 2017). Claudin-18 downregulation in gastric cancer has been linked with increased tumor size, invasion, and advanced tumor stage (Zhang et al. 2014). This reduction in claudin-18 has been associated with increased levels of miR-1303 in gastric cancer; miR-1303 binds to the 3′ untranslated region (UTR) of claudin-18, marking it for proteasomal degradation (Zhang et al. 2014). Studies of gastric cancer patient datasets have also identified that one of the most common genetic alterations of claudin-18 is the fusion of claudin-18 with Rho GTPase–activating protein 26 (CLDN18-ARHGAP), which is commonly found in metastatic diffuse-type gastric cancer (Tanaka et al. 2018; Li et al. 2020). Similarly, to claudin-11 and -18, claudin-23 expression was also found to be negatively associated with overall survival (Lu et al. 2017). Claudin-18.2 is a splice isoform of claudin-18 that is exclusively expressed in gastric tissue. Since its expression is maintained at high levels in gastric cancer cells, it is currently under active investigation for targeted therapy (Zhang et al. 2020).

Claudins are not the only TJ proteins that contribute to gastric cancer invasiveness. JAM-A can promote proliferation and invasion and inhibit apoptosis in gastric cancer independently (Ikeo et al. 2015). Ikeo et al. showed that knockout of JAM-A in rat and human gastric cancer cells, but not in normal gastric mucosal-derived cells, reduced proliferation, invasion, and expression of the anti-apoptotic protein B-cell lymphoma-extra large (Bcl-xL) (Ikeo et al. 2015). Tricellulin, a foundational TJ protein in tricellular junctions, is regulated by the EMT transcription factor Snail in gastric cancer. Snail overexpression in the HSC-45 gastric cancer cell line results in reduced tricellulin mRNA levels and membrane localization, as well as lower E-cadherin and increased vimentin and N-cadherin mRNA (Masuda et al. 2010).

1.8.4 Hepatocellular Carcinoma and Cholangiocarcinoma

In the liver, TJs between hepatocytes create the blood-biliary barrier, regulate bile canaliculi barrier function, and maintain hepatocyte apicobasal polarity (Gissen and Arias 2015; Rao and Samak 2013). As within other tissues and organ systems, TJ dysfunction is also commonly seen with hepatobiliary pathologies. For example, several studies have revealed increased claudin-1 expression on basolateral and apical hepatocyte membranes in both liver cirrhosis and hepatocellular carcinoma (HCC). Loss of claudin-1 expression in resected HCC was shown to be associated with more poorly differentiated HCC, higher degree of invasiveness, and poor patient outcomes (Holczbauer et al. 2014). In another study of surgically resected hepatoblastoma samples, claudin-1 and -2 expression were found to be significantly increased in samples with a differentiated fetal cell component compared to those with an embryonal component, and both claudins were inversely correlated with cell proliferation. Loss of claudin-1 and -2 in this liver cancer subtype was shown to be associated with the more aggressive embryonal phenotype (Halasz et al. 2006). Claudin-3, which inhibits Wnt-β-catenin signaling and EMT, was found to be downregulated or epigenetically silenced by promoter methylation in HCC. Further, in a study of 114 HCC patients, reduced expression of claudin-3 was found to be independently associated with lower overall survival (Jiang et al. 2014). Similarly, epigenetic silencing of claudin-14 has been linked to increased HCC aggressiveness (Li et al. 2016). Claudin-4 and claudin-7 expression was shown to be higher in cholangiocarcinoma compared to HCC (Ono et al. 2016). Precancerous biliary lesions and cholangiocarcinoma have been shown to express high levels of claudin-4. In two cholangiocarcinoma cell lines, claudin-4 targeting by siRNA reduced cell migration and invasion, but not proliferation. Higher claudin-7 expression in a small subset of HCC patients showed a tendency toward improved survival compared to low claudin-7 expression (Brokalaki et al. 2012). Elevated claudin-10 expression in HCC is associated with an increased incidence of tumor recurrence and poor outcomes following resection. Silencing claudin-10 in HLE hepatoma cells with high claudin-10 reduced tumorigenic potential and cell invasion (Ip et al. 2007). Downregulation of claudin-11 by miR-99b may promote HCC metastatic growth (Yang et al. 2015). Elevated serum levels of ZO-1 have been identified in HCC patients and are associated with disease severity using Child-Pugh class HCC staging. Interestingly, in this study, serum ZO-1 levels were significantly associated with the inflammatory marker c-reactive protein (CRP) (Ram et al. 2018).

1.8.5 Colorectal Cancer

Fluctuations in the expression of TJ proteins in colorectal cancer (CRC) point to specific changes during CRC progression from initial oncogenic transformation to invasion and metastasis. Claudin-2 expression is elevated in CRC in part due to EGF secreted in the tumor microenvironment and is positively associated with tumor grade (Dhawan et al. 2011). Compared to normal colon tissue, the expression of claudin-3 and claudin-4 is slightly elevated in membrane/cytoskeleton-associated fractions (1.5- and 2.5-fold, respectively) in tumor tissue, while claudin-1 is more robustly elevated in both cytosolic and membrane/cytoskeleton-associated fractions (5.7-fold) (de Oliveira et al. 2005). Claudin-1 expression is significantly elevated in CRC at both the mRNA and protein level (Caruso et al. 2014). Interestingly, claudin-7 is reduced in CRC, and its expression follows a reciprocal pattern to that of claudin-1 (Gowrikumar et al. 2021). Elevated claudin-1 expression and decreased claudin-7 expression were found to be associated with non-responsiveness to first-line chemotherapy for CRC (Gowrikumar et al. 2021). In the normal colon, as colon epithelial cells differentiate and migrate from the base of the colon crypt to the luminal surface, claudin-7 expression rises, forming an expression gradient from the basal surface to the lumen (Farkas et al. 2015). However, in CRC tissue, claudin-7 is significantly reduced, and, where present, aberrantly localized in the cytoplasm compared to surrounding nontumor tissue (Wang et al. 2018). In CRC cell lines, claudin-7 overexpression has been shown to confer growth inhibition, increase apoptosis, and promote a mesenchymal-to-epithelial transition. Conversely, the loss of claudin-7 in CRC cells was shown to increase the expression of EMT transcription factors and cell proliferation (Bhat et al. 2015). In this same study, claudin-7 overexpression in mouse tumor xenografts inhibited tumor growth. Cluster analysis of gene expression in human CRC samples indicated that claudin-7 overexpression was associated with increased levels of the growth factor bone morphogenetic protein 2 (BMP-2), Rab25, and CD55; further investigation showed that claudin-7 overexpression in SW620 cells was accompanied by an increase in Rab25 expression, as well as a reduction in p-ERK and p-Src. Analysis of the claudin-7 promoter region noted hypermethylation of CpG sites, mediated by DNA methyltransferase 3 apha (DNMT3a) (Bhat et al. 2015).

In one study, occludin expression was found to be inversely associated with CRC severity. One study analyzed occludin expression in 129 CRC specimens using immunohistochemistry and found that in 55 of 129 (42%) of tumors, occludin expression was absent or low compared to normal mucosa, while 71 of 129 (55%) samples revealed normal or increased occludin expression (Resnick et al. 2005). Statistical analysis of the data indicated a significant inverse relationship between occludin expression and tumor grade; however, additional studies are needed to confirm these findings.

The regulation of claudin-8 in CRC is also unclear. One study found claudin-8 to be elevated in CRC compared to normal adjacent tissue, and claudin-8-overexpression in CRC cell lines activated MAPK/ERK and increased the expression of MMP-9 (Cheng et al. 2019). Another study comparing TJ protein expression in CRC patient tissue with matched normal controls found strong claudin-8-expression in normal tissues, while tumor tissues displayed a slightly lower signal intensity, though these differences were not statistically significant (Grone et al. 2007).

Relatively few studies have examined the roles of JAMs and tricellulin in CRC. It is known that JAM-A is intricately involved in maintaining colon homeostasis, but investigation of its expression in patient samples is required to establish its association in CRC (Liu et al. 2000). On the other hand, tricellulin has been identified as the potential site of macromolecular penetration through the epithelial barrier, though how this relates to CRC pathogenesis or progression remains poorly understood (Krug et al. 2009).

1.8.6 IBD-Associated Cancer

It is well-established that chronic inflammation promotes tumorigenesis. Chronic inflammation can also affect TJ protein expression, which may further promote oncogenic transformation through a number of signaling pathways. Researchers found that both adherens junction and TJ proteins were expressed in patients with CRC and UC. Similarly to CRC, claudin-1 and -2 levels are elevated in IBD-associated carcinoma; both were found to be increased in IBD-associated dysplasia in comparison to nondysplastic IBD (Weber et al. 2008). Studies have also found elevated levels of claudin-1 and -2 in IBD patients with cancer. In the chronically inflamed or IBD colon, the presence of proinflammatory TNF-α contributes to claudin-1 upregulation. In turn, claudin-1 triggers the activation of the pro-tumorigenic ERK and Src signaling pathways (Bhat et al. 2016). In mouse models, claudin-1 overexpression in the colon causes exacerbated colitis and impairs recovery from experimental colitis due to its modulation of the Notch signaling pathway. When subjected to acute colitis for 7 days followed by a 10-day recovery period, claudin-1-transgenic mice showed significantly impaired recovery with a histologic predominance of dysplastic colon crypts compared to regenerative crypts. In a mouse model of colitis-associated cancer, claudin-1-transgenic mice exhibited an increased tumor burden compared to controls (Gowrikumar et al. 2019). Further, the dysplastic regions of claudin-1-transgenic mice from both experiments showed increased nuclear localization of Ser552-phosphorylated β-catenin, suggesting that nuclear localization of p-β-catenin^{Ser552} may contribute to the effects of claudin-1 on tumorigenesis in the setting of colitis (Gowrikumar et al. 2019).

Claudin-2, one of the pore-forming claudins, has shown to be elevated in colitis and is associated with the severity of inflammatory disease. Claudin-2 is upregulated by Cdx-associated Wnt signaling activation and contributes to tumorigenicity of colitis-associated cancer by promoting cell proliferation via EGFR/ERK signaling in vitro (Dhawan et al. 2011; Mankertz et al. 2004). However, knockout of claudin-2 in several experimental models of colitis has also been shown to worsen the effects of colitis; this is thought to be the result of claudin-2-associated increase in IL-13, which further increased epithelial barrier permeability and contributed to progression of immune-mediated colitis (Barrett 2020).

In a claudin-3-deficient mouse model, colonic epithelial dedifferentiation, barrier dysfunction, and rapid tumor progression were observed, likely as a result of IL-6/gp130/STAT3 signaling-induced upregulation of Wnt/β-catenin signaling. These results point to claudin-3 as a mediator of cross-talk between the IL-6 and the Wnt signaling pathways (Ahmad et al. 2017b). Further, claudin-7 inhibits ERK and Src signaling, thereby suppressing EMT and tumor progression in colorectal cancer cell lines (Bhat et al. 2015). However, according to one study, occludin expression and localization in colon dysplasia, acute self-limited colitis, and sporadic adenomas was comparable to controls (Weber et al. 2008) from TJ dysregulation in the setting of or resulting from autoinflammatory insults, as seen in chronic colitis or IBD, may be an important factor in the development of IBD-associated cancer; more studies are needed to determine how and whether TJs may be used as biomarkers or therapeutic targets to treat IBD and prevent neoplastic transformation.

1.9 Targeting Tight Junction Proteins in Inflammation and Cancer

The TJ complex presents an exciting new host of targets for the detection and treatment of a variety of pathologies. To target TJ protein alterations in inflammatory disease and carcinomas, novel and innovative approaches using monoclonal antibodies (mAbs), enterotoxins, and therapeutic gene delivery are currently being developed.

1.9.1 Immunotherapies

The use of immunotherapeutic agents such as monoclonal antibodies (mAbs) for targeting disease-specific antigens has increased over the past decade, owing in part to their target specificity and stability (Mori et al. 2007). mAbs generated against clinically relevant cell surface molecules can disrupt ligand binding and downstream signaling pathways. In addition, the presence of bound mAbs can trigger the recruitment and activation of immune cells (Weiner et al. 2010). Numerous investigations have evaluated the feasibility and efficacy of targeting TJ proteins with mAbs for the treatment of cancer. Of the TJ proteins, there are currently therapeutic antibodies developed against claudin-1, claudin-2, claudin-3, claudin-4, claudin-6, claudin-18.2, JAM-A, and JAM-C (Leech et al. 2015).

Zolbetuximab (IMAB362; formerly claudiximab) is a chimeric monoclonal IgG1 antibody generated against claudin-18.2. It has been tested in multiple clinical trials for the treatment of gastric cancer as a monotherapy, in combination with standard-of-care therapy with epirubicin, oxaliplatin, and capecitabine (EOX), and as an immunomodulatory agent in cancer immunotherapy (Sahin et al. 2018, 2021; Singh et al. 2017). In the MONO-2013 repeated-dose monotherapy trial, the response rate was found to be 10% and the disease control rate was 30% (Tureci et al. 2019). When administered high-dose in combination with EOX or as an immunomodulatory therapy in combination with zoledronic acid and IL-2, it was found to increase overall survival by 13.4 months and 40 weeks, respectively (Zhang et al. 2020; Sahin et al. 2015).

Experimental mAbs against claudin-1 have shown promise in experimental models. Anti-claudin-1 antibodies were the first investigated as a treatment for hepatitis C virus (HCV) infection (Zeisel et al. 2015). Interestingly, anti-claudin-1 antibodies have not shown significant adverse or off-target effects in a wide range of animal and human cell-based models (Mailly et al. 2015; Colpitts et al. 2018; Fofana et al. 2012). This is most likely because these anti-claudin-1 antibodies target the non-junctionally expressed claudin-1 without influencing TJ barrier function, as demonstrated in various TJ model systems (Mailly et al. 2015; Colpitts et al. 2018; Fofana et al. 2012). The claudin-1 antibody (6F6) produced by Cherradi et al. produced a claudin-1 antibody (6F6) has been shown to reduce primary and have shown that it reduces tumor development and liver metastasis in mouse models of CRC (Cherradi et al. 2017).

Other TJ proteins are also being investigated as treatment targets. Hashimoto et al. demonstrated that an anti-claudin-2 antibody can slow mouse xenograft tumor growth with minimal effects on mouse weight, kidney function, and liver function (Hashimoto et al. 2018). Targeting claudin-4 has been promising in treating ovarian and pancreatic cancers in vitro (Suzuki et al. 2009). A study utilizing a novel dual-targeting monoclonal antibody against both claudin-3 and claudin-4 reported a dramatic reduction in the growth of breast and cancer cell lines and in breast and ovarian tumor growth in vivo (Kato-Nakano et al. 2010). An anti-JAM-C antibody was effective in reducing tumor growth by decreasing tumor vascularization (Lamagna et al. 2005), and an antibody against JAM-A has also been shown to inhibit tumor growth in several mouse models (Goetsch et al. 2013).

Despite encouraging results and good tolerability, mAb therapy has several potential drawbacks. Large-scale production of biologic medications is expensive and technically challenging, which translates to higher medication costs for patients. Further, the need for high therapeutic antibody doses administered over an extended period of time for maximum efficacy may make the treatment regimen even less affordable. In addition, responses to therapy may vary depending on mAb stability and biodistribution, individual variability (age, gender, other diseases), and safety due to immunogenicity.

1.9.2 *Clostridium perfringens* Enterotoxin

Clostridium perfringens (*C. perfringens*) enterotoxin (CPE)-mediated therapy presents a unique approach to cancer treatment that benefits from specific expression patterns of claudin-3 and -4 in cancer cells. CPE is a 35-kDa polypeptide toxin produced by the bacterium *C. perfringens* that, if ingested orally via contaminated food products, causes nausea, vomiting, abdominal cramping, fever, and diarrhea.

Another known characteristic of CPE is its ability to rapidly lyse cells upon binding to its receptors, claudin-3 and claudin-4, directly affecting membrane permeability (Sarker et al. 1999). Given that claudin-3 and claudin-4 are overexpressed in numerous cancers, CPE presents an elegant, targeted approach to induce cell death specifically in cancer cells. One study carried out with CPE-mediated therapy resulted in rapid and dose-dependent cytolysis exclusively in breast cancer cells that express claudin-3 and claudin-4 and reduced tumor volume in xenograft mouse models (Kominsky et al. 2004). Similarly, pancreatic, prostate, ovarian, and brain tumors showed sensitivity to CPE treatment in mouse models (Santin et al. 2005; Michl et al. 2001; Long et al. 2001; Kominsky et al. 2007). Claudin-6 has also recently been identified as a binding target of CPE; binding assays reveal that CPE can bind claudin-6 in cells to induce cell cytotoxicity (Lal-Nag et al. 2012). Knock-out of claudin-6 decreased CPE sensitivity in the ovarian cancer cell line UCI-101, and overexpression of claudin-6 in CPE-resistant ovarian cancer cell lines sensitized them to CPE treatment (Lal-Nag et al. 2012). Although studies in mice have not demonstrated significant toxicity following intratumoral treatment with CPE, it must be noted that claudin-3 and -4 are also widely expressed in normal tissues as well as in cancers. This may pose as a risk in the systemic utilization of this therapy, potentially restricting it to cancers accessible via local application. In turn, this could lead to distribution problems within tumor tissue, principally in large solid tumor masses. The dependence of the expression of claudins and the need for multiple applications leading to immuno-sensitization against the treatment are other limitations of CPE-mediated therapy (Santin et al. 2005; Morin 2005).

1.9.3 DZ-50

Hensley et al. have found that quinazoline-α_1 adrenoreceptor antagonists commonly used to treat benign prostatic hyperplasia may also be an effective treatment modality for prostate cancer, given their induction of the extrinsic, caspase-8-dependent apoptosis pathway (Hensley et al. 2014). In an effort to identify pharmacotherapeutic strategies that exploit cell death mechanisms that cancer cells do not typically subvert, they have designed a similar drug known as DZ-50 that impairs tumor growth and metastasis by stimulating anoikis. Cancer cell resistance to anoikis driven by aberrant signaling sustained by the tumor microenvironment confers high invasive potential and therapeutic resistance. Using the human prostate cancer cell line DU-145, the authors found that DZ-50 also downregulated a number of genes including claudin-11 (Hensley et al. 2014). Confocal microscopy demonstrated structural disruption of both focal adhesions and TJs by the downregulation of these gene targets, leading to decreased prostate cancer cell survival, migration, and adhesion to extracellular matrix components in two androgen-independent human prostate cancer cell lines (Hensley et al. 2014). These findings suggest that DZ-50 exerts its antitumor effect by targeting focal adhesions and TJs to promote anoikis, supporting the therapeutic significance and further study of this agent for the treatment of advanced prostate cancer (Schulzke et al. 2005).

1.9.4 Glyburide

Inhibition of sulfonylurea receptor 1 (SUR1) by glyburide has been shown to decrease edema after subarachnoid hemorrhage, likely secondary to glyburide-induced changes in the localization of junctional ZO-1. SUR1 expression is also known to be increased in certain tumor subtypes, including non-small lung carcinoma (Xu et al. 2019) and glioblastoma (Thompson et al. 2018). Thompson et al. investigated whether inhibiting SUR1 reduced cerebral edema that occurs secondary to brain metastasis and explored whether SUR1 associated with TJ protein ZO-1 (Thompson et al. 2013). In an animal model of small cell lung carcinoma, glyburide plus dexamethasone decreased blood–tumor barrier permeability. In rat models of melanoma with brain metastases, glyburide alone inhibited the increased blood–tumor barrier permeability (Thompson et al. 2013), while dexamethasone alone only modestly lowered blood–tumor barrier permeability increase (Thompson et al. 2013). Administration of both glyburide and dexamethasone decreased the distance between ZO-1 proteins along the cell membrane, termed ZO-1 gaps. By decreasing ZO-1 gaps, glyburide was at least as effective as dexamethasone at halting increased blood–tumor barrier permeability caused by small cell lung carcinoma and melanoma. The authors concluded that glyburide is a safe, inexpensive, and efficacious alternative to dexamethasone for the treatment of cerebral metastasis-related vasogenic edema (Thompson et al. 2013).

1.9.5 Spantide III

Rodriguez et al. examined the role of the proinflammatory peptide substance P (SP) in the trafficking of breast cancer cells across the blood-brain barrier (BBB) using in vitro and in vivo models (Rodriguez et al. 2014). SP is secreted by breast cancer cells, and it enables breast cancer cells to attach and migrate across human brain microvascular endothelial cells in vitro. SP also stimulates brain microvascular endothelial cells to secrete TNF-α and angiotensin-2 (Ang-2), which are known to alter the localization and distribution of TJ proteins to affect BBB permeability. In this study, substance P reduced the expression of ZO-1 and claudin-5 and resulted in the disruption of cytoskeletal and TJ structures. Administration of the SP antagonist spantide III abrogated these effects in vivo and reduced the colonization and formation of brain metastases. Given these findings, SP inhibition may be a promising modality for preventing BBB alteration and brain metastases in advanced cancers (Rodriguez et al. 2014).

1.9.6 Coxsackie and Adenovirus Receptor (CAR) Overexpression

Since the advent of gene therapy, it has become increasingly important to identify receptors that enable specific binding of gene therapy vectors to target cells. The coxsackie and adenovirus receptor (CAR) is a TJ protein to which group B coxsackieviruses (CVB) and adenoviruses (Ad) must bind in order to infect cells. Given that CAR is known to be overexpressed in numerous cancers, including breast cancer, some lung cancers, and endometrial adenocarcinoma, using CAR-targeting CVB and Ad vectors may increase the specificity of gene delivery and, therefore, the efficacy of gene therapy (Leech et al. 2015; Coyne and Bergelson 2005; Martin et al. 2005; Wang et al. 2006; Giaginis et al. 2008). However, many cancers do not overexpress CAR, and some show a reduction in CAR expression. For example, gliomas, bladder cancers, prostate cancers, rhabdomyosarcoma, colorectal cancers, ovarian cancers, non-small cell lung cancers, and breast cancer metastases have low levels of CAR expression, limiting efficient adenovirus infection in those cancer cells (Miller et al. 1998; Fuxe et al. 2003; Kim et al. 2002; Shayakhmetov et al. 2002; Li et al. 1999; Okegawa et al. 2000; Cripe et al. 2001; Fechner et al. 2000; Qin et al. 2003). Investigation of the molecular mechanisms regulating CAR expression in cancers have shown that CAR can be both induced and inhibited by the same molecular signaling pathways, as in the case of the MAPK/ERK pathway in esophageal cancer (Ma et al. 2012). The differential expression and regulation of CAR in normal and cancerous tissue subtypes may limit the use of Ad and CVB vectors in gene therapy delivery.

Although most clinical trials using CAR-targeting viral vectors for therapeutic gene delivery have demonstrated good safety and tolerability profiles, there remains a number of concerns regarding adverse events related to the viral vectors (Sibbald 2001). Increased scrutiny of gene therapy trials using viral vectors exposed numerous problems related to the purification of viral particles, measurement of vector concentration, selection of hosts, and subject selection criteria (Vorburger and Hunt 2002).

1.10 Conclusion

Although several TJ protein targets have been identified for the treatment of cancer and inflammatory conditions, more research is necessary to determine the efficacy, safety, and feasibility of targeting them for therapy. Along with identifying novel therapeutics or repurposing approved drugs for TJ targeting, it is necessary to expand our knowledge about the mechanistic interactions involving TJ proteins in both physiological and pathological contexts.

References

Achari C, Winslow S, Larsson C (2015) Down regulation of CLDND1 induces apoptosis in breast cancer cells. PLoS One 10(6):e0130300

Agarwal R, D'Souza T, Morin PJ (2005) Claudin-3 and claudin-4 expression in ovarian epithelial cells enhances invasion and is associated with increased matrix metalloproteinase-2 activity. Cancer Res 65(16):7378–7385

Aggarwal S, Suzuki T, Taylor WL, Bhargava A, Rao RK (2011) Contrasting effects of ERK on tight junction integrity in differentiated and under-differentiated Caco-2 cell monolayers. Biochem J 433(1):51–63

Ahmad R, Chaturvedi R, Olivares-Villagómez D, Habib T, Asim M, Shivesh P, Polk DB, Wilson KT, Washington MK, Van Kaer L, Dhawan P, Singh AB (2014) Targeted colonic claudin-2 expression renders resistance to epithelial injury, induces immune suppression, and protects from colitis. Mucosal Immunol 7(6):1340–1353

Ahmad R, Rah B, Bastola D, Dhawan P, Singh AB (2017a) Obesity-induces organ and tissue specific tight junction restructuring and barrier deregulation by Claudin switching. Sci Rep 7 (1):5125

Ahmad R, Kumar B, Chen Z, Chen X, Muller D, Lele SM, Washington MK, Batra SK, Dhawan P, Singh AB (2017b) Loss of claudin-3 expression induces IL6/gp130/Stat3 signaling to promote colon cancer malignancy by hyperactivating Wnt/beta-catenin signaling. Oncogene 36 (47):6592–6604

Aiello NM, Maddipati R, Norgard RJ, Balli D, Li J, Yuan S, Yamazoe T, Black T, Sahmoud A, Furth EE, Bar-Sagi D, Stanger BZ (2018) EMT subtype influences epithelial plasticity and mode of cell migration. Dev Cell 45(6):681–695 e684

Akasaka H, Sato F, Morohashi S, Wu Y, Liu Y, Kondo J, Odagiri H, Hakamada K, Kijima H (2010) Anti-apoptotic effect of claudin-1 in tamoxifen-treated human breast cancer MCF-7 cells. BMC Cancer 10:548

Alexandre MD, Jeansonne BG, Renegar RH, Tatum R, Chen YH (2007) The first extracellular domain of claudin-7 affects paracellular Cl- permeability. Biochem Biophys Res Commun 357 (1):87–91

Al-Sadi R, Khatib K, Guo S, Ye D, Youssef M, Ma T (2011) Occludin regulates macromolecule flux across the intestinal epithelial tight junction barrier. Am J Physiol Gastrointest Liver Physiol 300(6):G1054–G1064

Amasheh S, Meiri N, Gitter AH, Schoneberg T, Mankertz J, Schulzke JD, Fromm M (2002) Claudin-2 expression induces cation-selective channels in tight junctions of epithelial cells. J Cell Sci 115(Pt 24):4969–4976

Amasheh S, Schmidt T, Mahn M, Florian P, Mankertz J, Tavalali S, Gitter AH, Schulzke JD, Fromm M (2005) Contribution of claudin-5 to barrier properties in tight junctions of epithelial cells. Cell Tissue Res 321(1):89–96

Ananthakrishnan AN, Bernstein CN, Iliopoulos D, Macpherson A, Neurath MF, Ali RAR, Vavricka SR, Fiocchi C (2018) Environmental triggers in IBD: a review of progress and evidence. Nat Rev Gastroenterol Hepatol 15(1):39–49

Ando-Akatsuka Y, Yonemura S, Itoh M, Furuse M, Tsukita S (1999) Differential behavior of E-cadherin and occludin in their colocalization with ZO-1 during the establishment of epithelial cell polarity. J Cell Physiol 179(2):115–125

Andreeva AY, Krause E, Muller EC, Blasig IE, Utepbergenov DI (2001) Protein kinase C regulates the phosphorylation and cellular localization of occludin. J Biol Chem 276(42):38480–38486

Antal CE, Hudson AM, Kang E, Zanca C, Wirth C, Stephenson NL, Trotter EW, Gallegos LL, Miller CJ, Furnari FB, Hunter T, Brognard J, Newton AC (2015) Cancer-associated protein kinase C mutations reveal kinase's role as tumor suppressor. Cell 160(3):489–502

Arcangeli ML, Frontera V, Bardin F, Thomassin J, Chetaille B, Adams S, Adams RH, Aurrand-Lions M (2012) The junctional adhesion molecule-B regulates JAM-C-dependent melanoma cell metastasis. FEBS Lett 586(22):4046–4051

Arimura Y, Nagaishi K, Hosokawa M (2011) Dynamics of claudins expression in colitis and colitis-associated cancer in rat. Methods Mol Biol 762:409–425

Aurrand-Lions M, Lamagna C, Dangerfield JP, Wang S, Herrera P, Nourshargh S, Imhof BA (2005) Junctional adhesion molecule-C regulates the early influx of leukocytes into tissues during inflammation. J Immunol 174(10):6406–6415

Autenrieth DM, Baumgart DC (2012) Toxic megacolon. Inflamm Bowel Dis 18(3):584–591

Akizuki R, Shimobaba S, Matsunaga T, Endo S, Ikari A (2017) Claudin-5, -7, and -18 suppress proliferation mediated by inhibition of phosphorylation of Akt in human lung squamous cell carcinoma. Biochim Biophys Acta, Mol Cell Res 1864(2):293–302. https://doi.org/10.1016/j.bbamcr.2016.11.018. Epub 2016 Nov 21. PMID: 27884700.

Barrett KE (2020) Claudin-2 pore causes leak that breaches the dam in intestinal inflammation. J Clin Invest 130(10):5100–5101

Basuroy S, Sheth P, Kuppuswamy D, Balasubramanian S, Ray RM, Rao RK (2003) Expression of kinase-inactive c-Src delays oxidative stress-induced disassembly and accelerates calcium-mediated reassembly of tight junctions in the Caco-2 cell monolayer*. J Biol Chem 278 (14):11916–11924

Bhat AA, Sharma A, Pope J, Krishnan M, Washington MK, Singh AB, Dhawan P (2012) Caudal homeobox protein Cdx-2 cooperates with Wnt pathway to regulate claudin-1 expression in colon cancer cells. PLoS One 7(6):e37174

Bhat AA, Pope JL, Smith JJ, Ahmad R, Chen X, Washington MK, Beauchamp RD, Singh AB, Dhawan P (2015) Claudin-7 expression induces mesenchymal to epithelial transformation (MET) to inhibit colon tumorigenesis. Oncogene 34(35):4570–4580

Bhat AA, Ahmad R, Uppada SB, Singh AB, Dhawan P (2016) Claudin-1 promotes TNF-alpha-induced epithelial-mesenchymal transition and migration in colorectal adenocarcinoma cells. Exp Cell Res 349(1):119–127

Bhat AA, Uppada S, Achkar IW, Hashem S, Yadav SK, Shanmugakonar M, Al-Naemi HA, Haris M, Uddin S (2018) Tight junction proteins and signaling pathways in cancer and inflammation: a functional crosstalk. Front Physiol 9:1942

Borka K, Kaliszky P, Szabo E, Lotz G, Kupcsulik P, Schaff Z, Kiss A (2007) Claudin expression in pancreatic endocrine tumors as compared with ductal adenocarcinomas. Virchows Arch 450 (5):549–557

Brennan K, McSherry EA, Hudson L, Kay EW, Hill AD, Young LS, Hopkins AM (2013) Junctional adhesion molecule-A is co-expressed with HER2 in breast tumors and acts as a novel regulator of HER2 protein degradation and signaling. Oncogene 32(22):2799–2804

Brokalaki EI, Weber F, Sotiropoulos GC, Daoudaki M, Cicinnati VR, Beckebaum S (2012) Claudin-7 expression in hepatocellular carcinoma. Transplant Proc 44(9):2737–2740

Brüning A, Stickeler E, Diederich D, Walz L, Rohleder H, Friese K, Runnebaum IB (2005) Coxsackie and adenovirus receptor promotes adenocarcinoma cell survival and is expressionally activated after transition from preneoplastic precursor lesions to invasive adenocarcinomas. Clin Cancer Res 11(12):4316–4320

Buchert M, Papin M, Bonnans C, Darido C, Raye WS, Garambois V, Pélegrin A, Bourgaux JF, Pannequin J, Joubert D, Hollande F (2010) Symplekin promotes tumorigenicity by up-regulating claudin-2 expression. Proc Natl Acad Sci U S A 107(6):2628–2633

Bürgel N, Bojarski C, Mankertz J, Zeitz M, Fromm M, Schulzke JD (2002) Mechanisms of diarrhea in collagenous colitis. Gastroenterology 123(2):433–443

Burstein E, Fearon ER (2008) Colitis and cancer: a tale of inflammatory cells and their cytokines. J Clin Invest 118(2):464–467

Capaldo CT, Nusrat A (2009) Cytokine regulation of tight junctions. Biochim Biophys Acta 1788 (4):864–871

Capaldo CT, Farkas AE, Hilgarth RS, Krug SM, Wolf MF, Benedik JK, Fromm M, Koval M, Parkos C, Nusrat A (2014) Proinflammatory cytokine-induced tight junction remodeling through dynamic self-assembly of claudins. Mol Biol Cell 25(18):2710–2719

Cario E, Gerken G, Podolsky DK (2004) Toll-like receptor 2 enhances ZO-1-associated intestinal epithelial barrier integrity via protein kinase C. Gastroenterology 127(1):224–238

Caruso M, Fung KY, Moore J, Brierley GV, Cosgrove LJ, Thomas M, Cheetham G, Brook E, Fraser LM, Tin T, Tran H, Ruszkiewicz A (2014) Claudin-1 expression is elevated in colorectal cancer precursor lesions harboring the BRAF V600E mutation. Transl Oncol 7(4):456–463

Chaouche-Mazouni S, Copin MC, Lassalle P, Lebaili N, Cortot A, Scherpereel A (2013) M14K and M38K malignant pleural mesothelioma cell lines preserve the same claudin-based phenotype in vivo. In Vivo 27(2):227–232

Chelakkot C, Ghim J, Rajasekaran N, Choi JS, Kim JH, Jang MH, Shin YK, Suh PG, Ryu SH (2017) Intestinal epithelial cell-specific deletion of PLD2 alleviates DSS-induced Colitis by regulating occludin. Sci Rep 7(1):1573

Chelakkot C, Ghim J, Ryu SH (2018) Mechanisms regulating intestinal barrier integrity and its pathological implications. Exp Mol Med 50(8):1–9

Chen Z, Wang Q, Sun J, Gu A, Jin M, Shen Z, Qiu Z, Wang J, Wang X, Zhan Z, Li JW (2013) Expression of the coxsackie and adenovirus receptor in human lung cancers. Tumour Biol 34(1):17–24

Cheng B, Rong A, Zhou Q, Li W (2019) CLDN8 promotes colorectal cancer cell proliferation, migration, and invasion by activating MAPK/ERK signaling. Cancer Manag Res 11:3741–3751

Cherradi S, Ayrolles-Torro A, Vezzo-Vie N, Gueguinou N, Denis V, Combes E, Boissiere F, Busson M, Canterel-Thouennon L, Mollevi C, Pugniere M, Bibeau F, Ychou M, Martineau P, Gongora C, Del Rio M (2017) Antibody targeting of claudin-1 as a potential colorectal cancer therapy. J Exp Clin Cancer Res 36(1):89

Choi YL, Kim J, Kwon MJ, Choi JS, Kim TJ, Bae DS, Koh SS, In YH, Park YW, Kim SH, Ahn G, Shin YK (2007) Expression profile of tight junction protein claudin 3 and claudin 4 in ovarian serous adenocarcinoma with prognostic correlation. Histol Histopathol 22(11):1185–1195

Coëffier M, Gloro R, Boukhettala N, Aziz M, Lecleire S, Vandaele N, Antonietti M, Savoye G, Bôle-Feysot C, Déchelotte P, Reimund JM, Ducrotté P (2010) Increased proteasome-mediated degradation of occludin in irritable bowel syndrome. Am J Gastroenterol 105(5):1181–1188

Colpitts CC, Tawar RG, Mailly L, Thumann C, Heydmann L, Durand SC, Xiao F, Robinet E, Pessaux P, Zeisel MB, Baumert TF (2018) Humanisation of a claudin-1-specific monoclonal antibody for clinical prevention and cure of HCV infection without escape. Gut 67(4):736–745

Cornick S, Tawiah A, Chadee K (2015) Roles and regulation of the mucus barrier in the gut. Tissue Barriers 3(1–2):e982426

Coyne CB, Bergelson JM (2005) CAR: a virus receptor within the tight junction. Adv Drug Deliv Rev 57(6):869–882

Coyne CB, Vanhook MK, Gambling TM, Carson JL, Boucher RC, Johnson LG (2002) Regulation of airway tight junctions by proinflammatory cytokines. Mol Biol Cell 13(9):3218–3234

Cripe TP, Dunphy EJ, Holub AD, Saini A, Vasi NH, Mahller YY, Collins MH, Snyder JD, Krasnykh V, Curiel DT, Wickham TJ, DeGregori J, Bergelson JM, Currier MA (2001) Fiber knob modifications overcome low, heterogeneous expression of the coxsackievirus-adenovirus receptor that limits adenovirus gene transfer and oncolysis for human rhabdomyosarcoma cells. Cancer Res 61(7):2953–2960

Cummins PM (2012) Occludin: one protein, many forms. Mol Cell Biol 32(2):242–250

Cuzic S, Antolic M, Ognjenovic A, Stupin-Polancec D, Petrinic Grba A, Hrvacic B, Dominis Kramaric M, Musladin S, Pozgaj L, Zlatar I, Polancec D, Aralica G, Banic M, Urek M, Mijandrusic Sincic B, Cubranic A, Glojnaric I, Bosnar M, Erakovic Haber V (2021) Claudins: beyond tight junctions in human IBD and murine models. Front Pharmacol 12:682614

de Oliveira SS, de Oliveira IM, De Souza W, Morgado-Diaz JA (2005) Claudins upregulation in human colorectal cancer. FEBS Lett 579(27):6179–6185

de Souza WF, Fortunato-Miranda N, Robbs BK, de Araujo WM, de Freitas-Junior JC, Bastos LG, Viola JP, Morgado-Díaz JA (2013) Claudin-3 overexpression increases the malignant potential of colorectal cancer cells: roles of ERK1/2 and PI3K-Akt as modulators of EGFR signaling. PLoS One 8(9):e74994

Deluco B, Fourie KR, Simko OM, Wilson HL (2021) Localization of Claudin-3 and Claudin-4 within the small intestine of newborn piglets. Physiol Rep 9(3):e14717

Dhawan P, Singh AB, Deane NG, No Y, Shiou SR, Schmidt C, Neff J, Washington MK, Beauchamp RD (2005) Claudin-1 regulates cellular transformation and metastatic behavior in colon cancer. J Clin Invest 115(7):1765–1776

Dhawan P, Ahmad R, Chaturvedi R, Smith JJ, Midha R, Mittal MK, Krishnan M, Chen X, Eschrich S, Yeatman TJ, Harris RC, Washington MK, Wilson KT, Beauchamp RD, Singh AB (2011) Claudin-2 expression increases tumorigenicity of colon cancer cells: role of epidermal growth factor receptor activation. Oncogene 30(29):3234–3247

Ebnet K (2017) Junctional Adhesion Molecules (JAMs): cell adhesion receptors with pleiotropic functions in cell physiology and development. Physiol Rev 97(4):1529–1554

Elias BC, Suzuki T, Seth A, Giorgianni F, Kale G, Shen L, Turner JR, Naren A, Desiderio DM, Rao R (2009) Phosphorylation of Tyr-398 and Tyr-402 in occludin prevents its interaction with ZO-1 and destabilizes its assembly at the tight junctions. J Biol Chem 284(3):1559–1569

El-Rayes BF, Ali S, Philip PA, Sarkar FH (2008) Protein kinase C: a target for therapy in pancreatic cancer. Pancreas 36(4):346–352

Fang Y, Yang J, Zu G, Cong C, Liu S, Xue F, Ma S, Liu J, Sun Y, Sun M (2021) Junctional adhesion molecule-like protein promotes tumor progression and metastasis via p38 signaling pathway in gastric cancer. Front Oncol 11:565676

Fantini MC, Pallone F (2008) Cytokines: from gut inflammation to colorectal cancer. Curr Drug Targets 9(5):375–380

Farkas AE, Hilgarth RS, Capaldo CT, Gerner-Smidt C, Powell DR, Vertino PM, Koval M, Parkos CA, Nusrat A (2015) HNF4alpha regulates claudin-7 protein expression during intestinal epithelial differentiation. Am J Pathol 185(8):2206–2218

Fechner H, Wang X, Wang H, Jansen A, Pauschinger M, Scherubl H, Bergelson JM, Schultheiss HP, Poller W (2000) Trans-complementation of vector replication versus Coxsackie-adenovirus-receptor overexpression to improve transgene expression in poorly permissive cancer cells. Gene Ther 7(22):1954–1968

Felinski EA, Antonetti DA (2005) Glucocorticoid regulation of endothelial cell tight junction gene expression: novel treatments for diabetic retinopathy. Curr Eye Res 30(11):949–957

Fevr T, Robine S, Louvard D, Huelsken J (2007) Wnt/beta-catenin is essential for intestinal homeostasis and maintenance of intestinal stem cells. Mol Cell Biol 27(21):7551–7559

Fofana I, Fafi-Kremer S, Carolla P, Fauvelle C, Zahid MN, Turek M, Heydmann L, Cury K, Hayer J, Combet C, Cosset FL, Pietschmann T, Hiet MS, Bartenschlager R, Habersetzer F, Doffoel M, Keck ZY, Foung SKH, Zeisel MB, Stoll-Keller F, Baumert TF (2012) Mutations that alter use of hepatitis C virus cell entry factors mediate escape from neutralizing antibodies. Gastroenterology 143(1):223–233 e229

Forrest JC, Campbell JA, Schelling P, Stehle T, Dermody TS (2003) Structure-function analysis of reovirus binding to junctional adhesion molecule 1. Implications for the mechanism of reovirus attachment. J Biol Chem 278(48):48434–48444

Förster C, Kahles T, Kietz S, Drenckhahn D (2007) Dexamethasone induces the expression of metalloproteinase inhibitor TIMP-1 in the murine cerebral vascular endothelial cell line cEND. J Physiol 580(Pt.3):937–949

Fujita H, Chiba H, Yokozaki H, Sakai N, Sugimoto K, Wada T, Kojima T, Yamashita T, Sawada N (2006) Differential expression and subcellular localization of claudin-7, -8, -12, -13, and -15 along the mouse intestine. J Histochem Cytochem 54(8):933–944

Furuse M, Hata M, Furuse K, Yoshida Y, Haratake A, Sugitani Y, Noda T, Kubo A, Tsukita S (2002) Claudin-based tight junctions are crucial for the mammalian epidermal barrier: a lesson from claudin-1-deficient mice. J Cell Biol 156(6):1099–1111

Fuse C, Ishida Y, Hikita T, Asai T, Oku N (2007) Junctional adhesion molecule-C promotes metastatic potential of HT1080 human fibrosarcoma. J Biol Chem 282(11):8276–8283

Fuxe J, Liu L, Malin S, Philipson L, Collins VP, Pettersson RF (2003) Expression of the coxsackie and adenovirus receptor in human astrocytic tumors and xenografts. Int J Cancer 103 (6):723–729

Gao M, Li W, Wang H, Wang G (2013) The distinct expression patterns of claudin-10, -14, -17 and E-cadherin between adjacent non-neoplastic tissues and gastric cancer tissues. Diagn Pathol 8:205

Garcia-Hernandez V, Quiros M, Nusrat A (2017) Intestinal epithelial claudins: expression and regulation in homeostasis and inflammation. Ann N Y Acad Sci 1397(1):66–79

Genua F, Raghunathan V, Jenab M, Gallagher WM, Hughes DJ (2021) The role of gut barrier dysfunction and microbiome dysbiosis in colorectal cancer development. Front Oncol 11:626349

Ghislin S, Obino D, Middendorp S, Boggetto N, Alcaide-Loridan C, Deshayes F (2011) Junctional adhesion molecules are required for melanoma cell lines transendothelial migration in vitro. Pigment Cell Melanoma Res 24(3):504–511

Giaginis CT, Zarros AC, Papaefthymiou MA, Papadopouli AE, Sfiniadakis IK, Theocharis SE (2008) Coxsackievirus and adenovirus receptor expression in human endometrial adenocarcinoma: possible clinical implications. World J Surg Oncol 6:59

Giaginis C, Zarros A, Alexandrou P, Klijanienko J, Delladetsima I, Theocharis S (2010) Evaluation of coxsackievirus and adenovirus receptor expression in human benign and malignant thyroid lesions. APMIS 118(3):210–221

Gissen P, Arias IM (2015) Structural and functional hepatocyte polarity and liver disease. J Hepatol 63(4):1023–1037

Goetsch L, Haeuw JF, Beau-Larvor C, Gonzalez A, Zanna L, Malissard M, Lepecquet AM, Robert A, Bailly C, Broussas M, Corvaia N (2013) A novel role for junctional adhesion molecule-A in tumor proliferation: modulation by an anti-JAM-A monoclonal antibody. Int J Cancer 132 (6):1463–1474

González-Mariscal L, Tapia R, Chamorro D (2008) Crosstalk of tight junction components with signaling pathways. Biochim Biophys Acta 1778(3):729–756

González-Mariscal L, Domínguez-Calderón A, Raya-Sandino A, Ortega-Olvera JM, Vargas-Sierra O, Martínez-Revollar G (2014) Tight junctions and the regulation of gene expression. Semin Cell Dev Biol 36:213–223

Götte M, Mohr C, Koo CY, Stock C, Vaske AK, Viola M, Ibrahim SA, Peddibhotla S, Teng YH, Low JY, Ebnet K, Kiesel L, Yip GW (2010) miR-145-dependent targeting of junctional adhesion molecule A and modulation of fascin expression are associated with reduced breast cancer cell motility and invasiveness. Oncogene 29(50):6569–6580

Gowrikumar S, Ahmad R, Uppada SB, Washington MK, Shi C, Singh AB, Dhawan P (2019) Upregulated claudin-1 expression promotes colitis-associated cancer by promoting beta-catenin phosphorylation and activation in Notch/p-AKT-dependent manner. Oncogene 38(26):5321–5337

Gowrikumar S, Primeaux M, Pravoverov K, Wu C, Szeglin BC, Sauve CG, Thapa I, Bastola D, Chen XS, Smith JJ, Singh AB, Dhawan P (2021) A claudin-based molecular signature identifies high-risk, chemoresistant colorectal cancer patients. Cell 10(9)

Grone J, Weber B, Staub E, Heinze M, Klaman I, Pilarsky C, Hermann K, Castanos-Velez E, Ropcke S, Mann B, Rosenthal A, Buhr HJ (2007) Differential expression of genes encoding tight junction proteins in colorectal cancer: frequent dysregulation of claudin-1, -8 and -12. Int J Color Dis 22(6):651–659

Gruber L, Hemmerling J, Schüppel V, Müller M, Boekschoten MV, Haller D (2015) Maternal high-fat diet accelerates development of Crohn's disease-like Ileitis in TNFΔARE/WT offspring. Inflamm Bowel Dis 21(9):2016–2025

Günzel D, Yu ASL (2013) Claudins and the modulation of tight junction permeability. Physiol Rev 93(2):525–569

Guo S, Nighot M, Al-Sadi R, Alhmoud T, Nighot P, Ma TY (2015) Lipopolysaccharide regulation of intestinal tight junction permeability is mediated by TLR4 signal transduction pathway activation of FAK and MyD88. J Immunol 195(10):4999–5010

Halasz J, Holczbauer A, Paska C, Kovacs M, Benyo G, Verebely T, Schaff Z, Kiss A (2006) Claudin-1 and claudin-2 differentiate fetal and embryonal components in human hepatoblastoma. Hum Pathol 37(5):555–561

Hao S, Yang Y, Liu Y, Yang S, Wang G, Xiao J, Liu H (2014) JAM-C promotes lymphangiogenesis and nodal metastasis in non-small cell lung cancer. Tumour Biol 35 (6):5675–5687

Hashimoto Y, Hata T, Tada M, Iida M, Watari A, Okada Y, Doi T, Kuniyasu H, Yagi K, Kondoh M (2018) Safety evaluation of a human chimeric monoclonal antibody that recognizes the extracellular loop domain of claudin-2. Eur J Pharm Sci 117:161–167

Henao-Mejia J, Elinav E, Jin C, Hao L, Mehal WZ, Strowig T, Thaiss CA, Kau AL, Eisenbarth SC, Jurczak MJ, Camporez J-P, Shulman GI, Gordon JI, Hoffman HM, Flavell RA (2012) Inflammasome-mediated dysbiosis regulates progression of NAFLD and obesity. Nature 482 (7384):179–185

Hensley PJ, Desiniotis A, Wang C, Stromberg A, Chen CS, Kyprianou N (2014) Novel pharmacologic targeting of tight junctions and focal adhesions in prostate cancer cells. PLoS One 9(1): e86238

Hewitt KJ, Agarwal R, Morin PJ (2006) The claudin gene family: expression in normal and neoplastic tissues. BMC Cancer 6:186

Hintsala HR, Siponen M, Haapasaari KM, Karihtala P, Soini Y (2013) Claudins 1, 2, 3, 4, 5 and 7 in solar keratosis and squamocellular carcinoma of the skin. Int J Clin Exp Pathol 6(12):2855–2863

Holczbauer A, Gyongyosi B, Lotz G, Torzsok P, Kaposi-Novak P, Szijarto A, Tatrai P, Kupcsulik P, Schaff Z, Kiss A (2014) Increased expression of claudin-1 and claudin-7 in liver cirrhosis and hepatocellular carcinoma. Pathol Oncol Res 20(3):493–502

Hong L, Wu Y, Feng J, Yu S, Li C, Wu Y, Li Z, Cao L, Wang F, Zhang Y (2014) Overexpression of the cell adhesion molecule claudin-9 is associated with invasion in pituitary oncocytomas. Hum Pathol 45(12):2423–2429

Hou J, Renigunta A, Yang J, Waldegger S (2010) Claudin-4 forms paracellular chloride channel in the kidney and requires claudin-8 for tight junction localization. Proc Natl Acad Sci U S A 107 (42):18010–18015

Hu JE, Bojarski C, Branchi F, Fromm M, Krug SM (2020) Leptin downregulates Angulin-1 in active Crohn's disease via STAT3. Int J Mol Sci 21(21)

Huo Q, Kinugasa T, Wang L, Huang J, Zhao J, Shibaguchi H, Kuroki M, Tanaka T, Yamashita Y, Nabeshima K, Iwasaki H, Kuroki M (2009) Claudin-1 protein is a major factor involved in the tumorigenesis of colorectal cancer. Anticancer Res 29(3):851–857

Hwang TL, Changchien TT, Wang CC, Wu CM (2014) Claudin-4 expression in gastric cancer cells enhances the invasion and is associated with the increased level of matrix metalloproteinase-2 and -9 expression. Oncol Lett 8(3):1367–1371

Ikari A, Sato T, Watanabe R, Yamazaki Y, Sugatani J (2012) Increase in claudin-2 expression by an EGFR/MEK/ERK/c-Fos pathway in lung adenocarcinoma A549 cells. Biochim Biophys Acta 1823(6):1110–1118

Ikari A, Watanabe R, Sato T, Taga S, Shimobaba S, Yamaguchi M, Yamazaki Y, Endo S, Matsunaga T, Sugatani J (2014) Nuclear distribution of claudin-2 increases cell proliferation in human lung adenocarcinoma cells. Biochim Biophys Acta 1843(9):2079–2088

Ikeo K, Oshima T, Shan J, Matsui H, Tomita T, Fukui H, Watari J, Miwa H (2015) Junctional adhesion molecule-A promotes proliferation and inhibits apoptosis of gastric cancer. Hepato-Gastroenterology 62(138):540–545

Ip YC, Cheung ST, Lee YT, Ho JC, Fan ST (2007) Inhibition of hepatocellular carcinoma invasion by suppression of claudin-10 in HLE cells. Mol Cancer Ther 6(11):2858–2867

Ito M, Kodama M, Masuko M, Yamaura M, Fuse K, Uesugi Y, Hirono S, Okura Y, Kato K, Hotta Y, Honda T, Kuwano R, Aizawa Y (2000) Expression of coxsackievirus and adenovirus receptor in hearts of rats with experimental autoimmune myocarditis. Circ Res 86(3):275–280

Itoh M, Bissell MJ (2003) The organization of tight junctions in epithelia: implications for mammary gland biology and breast tumorigenesis. J Mammary Gland Biol Neoplasia 8 (4):449–462

Jain S, Suzuki T, Seth A, Samak G, Rao R (2011) Protein kinase Cζ phosphorylates occludin and promotes assembly of epithelial tight junctions. Biochem J 437(2):289–299

Jiang L, Yang YD, Fu L, Xu W, Liu D, Liang Q, Zhang X, Xu L, Guan XY, Wu B, Sung JJ, Yu J (2014) CLDN3 inhibits cancer aggressiveness via Wnt-EMT signaling and is a potential prognostic biomarker for hepatocellular carcinoma. Oncotarget 5(17):7663–7676

Jun KH, Kim JH, Jung JH, Choi HJ, Chin HM (2014) Expression of claudin-7 and loss of claudin-18 correlate with poor prognosis in gastric cancer. Int J Surg 12(2):156–162

Kale G, Naren AP, Sheth P, Rao RK (2003) Tyrosine phosphorylation of occludin attenuates its interactions with ZO-1, ZO-2, and ZO-3. Biochem Biophys Res Commun 302(2):324–329

Kato-Nakano M, Suzuki M, Kawamoto S, Furuya A, Ohta S, Nakamura K, Ando H (2010) Characterization and evaluation of the antitumour activity of a dual-targeting monoclonal antibody against claudin-3 and claudin-4. Anticancer Res 30(11):4555–4562

Kawasaki T, Kawai T (2014) Toll-like receptor signaling pathways. Front Immunol 5

Kim JS, Lee SH, Cho YS, Choi JJ, Kim YH, Lee JH (2002) Enhancement of the adenoviral sensitivity of human ovarian cancer cells by transient expression of coxsackievirus and adenovirus receptor (CAR). Gynecol Oncol 85(2):260–265

Kim SS, Choi YD, Jin XM, Cho YM, Jang JJ, Juhng SW, Choi C (2009) Immunohistochemical stain for cytokeratin 7, S100A1 and claudin 8 is valuable in differential diagnosis of chromophobe renal cell carcinoma from renal oncocytoma. Histopathology 54(5):633–635

Kimbung S, Kovács A, Bendahl PO, Malmström P, Fernö M, Hatschek T, Hedenfalk I (2014) Claudin-2 is an independent negative prognostic factor in breast cancer and specifically predicts early liver recurrences. Mol Oncol 8(1):119–128

Kimura Y, Shiozaki H, Hirao M, Maeno Y, Doki Y, Inoue M, Monden T, Ando-Akatsuka Y, Furuse M, Tsukita S, Monden M (1997) Expression of occludin, tight-junction-associated protein, in human digestive tract. Am J Pathol 151(1):45–54

Kinugasa T, Huo Q, Higashi D, Shibaguchi H, Kuroki M, Tanaka T, Futami K, Yamashita Y, Hachimine K, Maekawa S, Nabeshima K, Iwasaki H, Kuroki M (2007) Selective up-regulation of claudin-1 and claudin-2 in colorectal cancer. Anticancer Res 27(6a):3729–3734

Kojima T, Takasawa A, Kyuno D, Ito T, Yamaguchi H, Hirata K, Tsujiwaki M, Murata M, Tanaka S, Sawada N (2011) Downregulation of tight junction-associated MARVEL protein marvelD3 during epithelial-mesenchymal transition in human pancreatic cancer cells. Exp Cell Res 317 (16):2288–2298

Kojima T, Kyuno D, Sawada N (2012) Targeting claudin-4 in human pancreatic cancer. Expert Opin Ther Targets 16(9):881–887

Kojima T, Yamaguchi H, Ito T, Kyuno D, Kono T, Konno T, Sawada N (2013) Tight junctions in human pancreatic duct epithelial cells. Tissue Barriers 1(4):e24894

Kolokytha P, Yiannou P, Keramopoulos D, Kolokythas A, Nonni A, Patsouris E, Pavlakis K (2014) Claudin-3 and claudin-4: distinct prognostic significance in triple-negative and luminal breast cancer. Appl Immunohistochem Mol Morphol 22(2):125–131

Kominsky SL, Argani P, Korz D, Evron E, Raman V, Garrett E, Rein A, Sauter G, Kallioniemi OP, Sukumar S (2003) Loss of the tight junction protein claudin-7 correlates with histological grade in both ductal carcinoma in situ and invasive ductal carcinoma of the breast. Oncogene 22 (13):2021–2033

Kominsky SL, Vali M, Korz D, Gabig TG, Weitzman SA, Argani P, Sukumar S (2004) Clostridium perfringens enterotoxin elicits rapid and specific cytolysis of breast carcinoma cells mediated through tight junction proteins claudin 3 and 4. Am J Pathol 164(5):1627–1633

Kominsky SL, Tyler B, Sosnowski J, Brady K, Doucet M, Nell D, Smedley JG 3rd, McClane B, Brem H, Sukumar S (2007) Clostridium perfringens enterotoxin as a novel-targeted therapeutic for brain metastasis. Cancer Res 67(17):7977–7982

Kondo J, Sato F, Kusumi T, Liu Y, Motonari O, Sato T, Kijima H (2008) Claudin-1 expression is induced by tumor necrosis factor-alpha in human pancreatic cancer cells. Int J Mol Med 22 (5):645–649

Konecny GE, Agarwal R, Keeney GA, Winterhoff B, Jones MB, Mariani A, Riehle D, Neuper C, Dowdy SC, Wang HJ, Morin PJ, Podratz KC (2008) Claudin-3 and claudin-4 expression in serous papillary, clear-cell, and endometrioid endometrial cancer. Gynecol Oncol 109(2):263–269

Konrad M, Schaller A, Seelow D, Pandey AV, Waldegger S, Lesslauer A, Vitzthum H, Suzuki Y, Luk JM, Becker C, Schlingmann KP, Schmid M, Rodriguez-Soriano J, Ariceta G, Cano F, Enriquez R, Juppner H, Bakkaloglu SA, Hediger MA, Gallati S, Neuhauss SC, Nurnberg P, Weber S (2006) Mutations in the tight-junction gene claudin 19 (CLDN19) are associated with renal magnesium wasting, renal failure, and severe ocular involvement. Am J Hum Genet 79 (5):949–957

Kopp JL, Dubois CL, Schaeffer DF, Samani A, Taghizadeh F, Cowan RW, Rhim AD, Stiles BL, Valasek M, Sander M (2018) Loss of Pten and activation of Kras synergistically induce formation of intraductal papillary mucinous neoplasia from pancreatic ductal cells in mice. Gastroenterology 154(5):1509–1523 e1505

Korompay A, Borka K, Lotz G, Somoracz A, Torzsok P, Erdelyi-Belle B, Kenessey I, Baranyai Z, Zsoldos F, Kupcsulik P, Bodoky G, Schaff Z, Kiss A (2012) Tricellulin expression in normal and neoplastic human pancreas. Histopathology 60(6B):E76–E86

Krug SM, Amasheh S, Richter JF, Milatz S, Gunzel D, Westphal JK, Huber O, Schulzke JD, Fromm M (2009) Tricellulin forms a barrier to macromolecules in tricellular tight junctions without affecting ion permeability. Mol Biol Cell 20(16):3713–3724

Krug SM, Bojarski C, Fromm A, Lee IM, Dames P, Richter JF, Turner JR, Fromm M, Schulzke JD (2018) Tricellulin is regulated via interleukin-13-receptor alpha2, affects macromolecule uptake, and is decreased in ulcerative colitis. Mucosal Immunol 11(2):345–356

Kuo SJ, Chien SY, Lin C, Chan SE, Tsai HT, Chen DR (2010) Significant elevation of CLDN16 and HAPLN3 gene expression in human breast cancer. Oncol Rep 24(3):759–766

Kuo WT, Shen L, Zuo L, Shashikanth N, Ong M, Wu L, Zha J, Edelblum KL, Wang Y, Wang Y, Nilsen SP, Turner JR (2019) Inflammation-induced Occludin downregulation limits epithelial apoptosis by suppressing Caspase-3 expression. Gastroenterology 157(5):1323–1337

Kyuno D, Kojima T, Yamaguchi H, Ito T, Kimura Y, Imamura M, Takasawa A, Murata M, Tanaka S, Hirata K, Sawada N (2013) Protein kinase Calpha inhibitor protects against downregulation of claudin-1 during epithelial-mesenchymal transition of pancreatic cancer. Carcinogenesis 34 (6):1232–1243

Kyuno D, Zhao K, Bauer N, Ryschich E, Zoller M (2019) Therapeutic targeting cancer-initiating cell markers by exosome miRNA: efficacy and functional consequences exemplified for claudin7 and EpCAM. Transl Oncol 12(2):191–199

Kyuno D, Takasawa A, Kikuchi S, Takemasa I, Osanai M, Kojima T (2021) Role of tight junctions in the epithelial-to-mesenchymal transition of cancer cells. Biochim Biophys Acta Biomembr 1863(3):183503

Lal-Nag M, Battis M, Santin AD, Morin PJ (2012) Claudin-6: a novel receptor for CPE-mediated cytotoxicity in ovarian cancer. Oncogenesis 1:e33

Lamagna C, Hodivala-Dilke KM, Imhof BA, Aurrand-Lions M (2005) Antibody against junctional adhesion molecule-C inhibits angiogenesis and tumor growth. Cancer Res 65(13):5703–5710

Lampis A, Hahne JC, Gasparini P, Cascione L, Hedayat S, Vlachogiannis G, Murgia C, Fontana E, Edwards J, Horgan PG, Terracciano L, Sansom OJ, Martins CD, Kramer-Marek G, Croce CM, Braconi C, Fassan M, Valeri N (2021) MIR21-induced loss of junctional adhesion molecule A promotes activation of oncogenic pathways, progression and metastasis in colorectal cancer. Cell Death Differ 28(10):2970–2982

Landy J, Ronde E, English N, Clark SK, Hart AL, Knight SC, Ciclitira PJ, Al-Hassi HO (2016) Tight junctions in inflammatory bowel diseases and inflammatory bowel disease associated colorectal cancer. World J Gastroenterol 22(11):3117–3126

Langer HF, Orlova VV, Xie C, Kaul S, Schneider D, Lonsdorf AS, Fahrleitner M, Choi EY, Dutoit V, Pellegrini M, Grossklaus S, Nawroth PP, Baretton G, Santoso S, Hwang ST, Arnold B, Chavakis T (2011) A novel function of junctional adhesion molecule-C in mediating melanoma cell metastasis. Cancer Res 71(12):4096–4105

Laukoetter MG, Nava P, Lee WY, Severson EA, Capaldo CT, Babbin BA, Williams IR, Koval M, Peatman E, Campbell JA, Dermody TS, Nusrat A, Parkos CA (2007) JAM-A regulates permeability and inflammation in the intestine in vivo. J Exp Med 204(13):3067–3076

Le N, Mazahery C, Nguyen K, Levine AD (2021) Regulation of intestinal epithelial barrier and immune function by activated T cells. Cellular Mol Gastroenterol Hepatol 11(1):55–76

Lee JW, Lee SJ, Seo J, Song SY, Ahn G, Park CS, Lee JH, Kim BG, Bae DS (2005) Increased expressions of claudin-1 and claudin-7 during the progression of cervical neoplasia. Gynecol Oncol 97(1):53–59

Lee JH, Lee YK, Lim JJ, Byun HO, Park I, Kim GH, Xu WG, Wang HJ, Yoon G (2015) Mitochondrial respiratory dysfunction induces claudin-1 expression via reactive oxygen species-mediated heat shock factor 1 activation, leading to hepatoma cell invasiveness. J Biol Chem 290(35):21421–21431

Leech AO, Cruz RG, Hill AD, Hopkins AM (2015) Paradigms lost-an emerging role for over-expression of tight junction adhesion proteins in cancer pathogenesis. Ann Transl Med 3 (13):184

Leinster DA, Colom B, Whiteford JR, Ennis DP, Lockley M, McNeish IA, Aurrand-Lions M, Chavakis T, Imhof BA, Balkwill FR, Nourshargh S (2013) Endothelial cell junctional adhesion molecule C plays a key role in the development of tumors in a murine model of ovarian cancer. FASEB J 27(10):4244–4253

Li Y, Pong RC, Bergelson JM, Hall MC, Sagalowsky AI, Tseng CP, Wang Z, Hsieh JT (1999) Loss of adenoviral receptor expression in human bladder cancer cells: a potential impact on the efficacy of gene therapy. Cancer Res 59(2):325–330

Li GZ, Wang ZH, Cui W, Fu JL, Wang YR, Liu P (2012) Tumor necrosis factor alpha increases intestinal permeability in mice with fulminant hepatic failure. World J Gastroenterol 18 (36):5042–5050

Li CP, Cai MY, Jiang LJ, Mai SJ, Chen JW, Wang FW, Liao YJ, Chen WH, Jin XH, Pei XQ, Guan XY, Zeng MS, Xie D (2016) CLDN14 is epigenetically silenced by EZH2-mediated H3K27ME3 and is a novel prognostic biomarker in hepatocellular carcinoma. Carcinogenesis 37(6):557–566

Li J, Zhang Y, Hu D, Gong T, Xu R, Gao J (2020) Analysis of the expression and genetic alteration of CLDN18 in gastric cancer. Aging (Albany NY) 12(14):14271–14284

Lin Z, Zhang X, Liu Z, Liu Q, Wang L, Lu Y, Liu Y, Wang M, Yang M, Jin X, Quan C (2013) The distinct expression patterns of claudin-2, -6, and -11 between human gastric neoplasms and adjacent non-neoplastic tissues. Diagn Pathol 8:133

Lipschutz JH, Li S, Arisco A, Balkovetz DF (2005) Extracellular signal-regulated kinases 1/ 2 control Claudin-2 expression in Madin-Darby canine kidney strain I and II cells. J Biol Chem 280(5):3780–3788

Liu Y, Nusrat A, Schnell FJ, Reaves TA, Walsh S, Pochet M, Parkos CA (2000) Human junction adhesion molecule regulates tight junction resealing in epithelia. J Cell Sci 113(Pt 13):2363–2374

Liu Y, Wang L, Lin XY, Wang J, Yu JH, Miao Y, Wang EH (2012) Anti-apoptotic effect of claudin-1 on TNF-α-induced apoptosis in human breast cancer MCF-7 cells. Tumour Biol 33 (6):2307–2315

Liu LB, Liu XB, Ma J, Liu YH, Li ZQ, Ma T, Zhao XH, Xi Z, Xue YX (2015) Bradykinin increased the permeability of BTB via NOS/NO/ZONAB-mediating down-regulation of claudin-5 and occludin. Biochem Biophys Res Commun 464(1):118–125

Loddo I, Romano C (2015) Inflammatory bowel disease: genetics, epigenetics, and pathogenesis. Front Immunol 6:551

Long H, Crean CD, Lee WH, Cummings OW, Gabig TG (2001) Expression of Clostridium perfringens enterotoxin receptors claudin-3 and claudin-4 in prostate cancer epithelium. Cancer Res 61(21):7878–7881

Lu Z, Ding L, Lu Q, Chen YH (2013) Claudins in intestines: distribution and functional significance in health and diseases. Tissue Barriers 1(3):e24978

Lu Y, Jing J, Sun L, Gong Y, Chen M, Wang Z, Sun M, Yuan Y (2017) Expression of claudin-11, -23 in different gastric tissues and its relationship with the risk and prognosis of gastric cancer. PLoS One 12(3):e0174476

Luettig J, Rosenthal R, Barmeyer C, Schulzke JD (2015) Claudin-2 as a mediator of leaky gut barrier during intestinal inflammation. Tissue Barriers 3(1–2):e977176

Luissint AC, Nusrat A, Parkos CA (2014) JAM-related proteins in mucosal homeostasis and inflammation. Semin Immunopathol 36(2):211–226

Lv J, Sun B, Mai Z, Jiang M, Du J (2017) CLDN-1 promoted the epithelial to migration and mesenchymal transition (EMT) in human bronchial epithelial cells via Notch pathway. Mol Cell Biochem 432(1–2):91–98

Ma J, Zhao J, Lu J, Jiang Y, Yang H, Li P, Zhao M, Liu K, Dong Z (2012) Coxsackievirus and adenovirus receptor promotes antitumor activity of oncolytic adenovirus H101 in esophageal cancer. Int J Mol Med 30(6):1403–1409

Ma J, Yao Y, Wang P, Liu Y, Zhao L, Li Z, Li Z, Xue Y (2014) MiR-181a regulates blood-tumor barrier permeability by targeting Krüppel-like factor 6. J Cereb Blood Flow Metab 34 (11):1826–1836

Maekawa Y, Ouzounian M, Opavsky MA, Liu PP (2007) Connecting the missing link between dilated cardiomyopathy and viral myocarditis: virus, cytoskeleton, and innate immunity. Circulation 115(1):5–8

Mailly L, Xiao F, Lupberger J, Wilson GK, Aubert P, Duong FHT, Calabrese D, Leboeuf C, Fofana I, Thumann C, Bandiera S, Lutgehetmann M, Volz T, Davis C, Harris HJ, Mee CJ, Girardi E, Chane-Woon-Ming B, Ericsson M, Fletcher N, Bartenschlager R, Pessaux P, Vercauteren K, Meuleman P, Villa P, Kaderali L, Pfeffer S, Heim MH, Neunlist M, Zeisel MB, Dandri M, McKeating JA, Robinet E, Baumert TF (2015) Clearance of persistent hepatitis C virus infection in humanized mice using a claudin-1-targeting monoclonal antibody. Nat Biotechnol 33 (5):549–554

Mandel I, Paperna T, Glass-Marmor L, Volkowich A, Badarny S, Schwartz I, Vardi P, Koren I, Miller A (2012) Tight junction proteins expression and modulation in immune cells and multiple sclerosis. J Cell Mol Med 16(4):765–775

Mankertz J, Hillenbrand B, Tavalali S, Huber O, Fromm M, Schulzke JD (2004) Functional crosstalk between Wnt signaling and Cdx-related transcriptional activation in the regulation of the claudin-2 promoter activity. Biochem Biophys Res Commun 314(4):1001–1007

Marchiando AM, Shen L, Graham WV, Weber CR, Schwarz BT, Austin JR 2nd, Raleigh DR, Guan Y, Watson AJ, Montrose MH, Turner JR (2010) Caveolin-1-dependent occludin endocytosis is required for TNF-induced tight junction regulation in vivo. J Cell Biol 189(1):111–126

Martin TA, Jiang WG (2009) Loss of tight junction barrier function and its role in cancer metastasis. Biochim Biophys Acta 1788(4):872–891

Martin TA, Watkins G, Jiang WG (2005) The Coxsackie-adenovirus receptor has elevated expression in human breast cancer. Clin Exp Med 5(3):122–128

Martin TA, Mansel RE, Jiang WG (2010) Loss of occludin leads to the progression of human breast cancer. Int J Mol Med 26(5):723–734

Martin TA, Lane J, Ozupek H, Jiang WG (2013) Claudin-20 promotes an aggressive phenotype in human breast cancer cells. Tissue Barriers 1(3):e26518

Masuda R, Semba S, Mizuuchi E, Yanagihara K, Yokozaki H (2010) Negative regulation of the tight junction protein tricellulin by snail-induced epithelial-mesenchymal transition in gastric carcinoma cells. Pathobiology 77(2):106–113

McClane BA (2001) The complex interactions between Clostridium perfringens enterotoxin and epithelial tight junctions. Toxicon 39(11):1781–1791

McSherry EA, McGee SF, Jirstrom K, Doyle EM, Brennan DJ, Landberg G, Dervan PA, Hopkins AM, Gallagher WM (2009) JAM-A expression positively correlates with poor prognosis in breast cancer patients. Int J Cancer 125(6):1343–1351

McSherry EA, Brennan K, Hudson L, Hill AD, Hopkins AM (2011) Breast cancer cell migration is regulated through junctional adhesion molecule-A-mediated activation of Rap1 GTPase. Breast Cancer Res 13(2):R31

Men W, Martin TA, Ruge F, Zhang N, Du P, Yang Y, Jiang WG (2015) Expression of claudins in human clear cell renal cell carcinoma. Cancer Genomics Proteomics 12(1):1–8

Miao YS, Zhao YY, Zhao LN, Wang P, Liu YH, Ma J, Xue YX (2015) MiR-18a increased the permeability of BTB via RUNX1 mediated down-regulation of ZO-1, occludin and claudin-5. Cell Signal 27(1):156–167

Michl P, Buchholz M, Rolke M, Kunsch S, Lohr M, McClane B, Tsukita S, Leder G, Adler G, Gress TM (2001) Claudin-4: a new target for pancreatic cancer treatment using Clostridium perfringens enterotoxin. Gastroenterology 121(3):678–684

Michl P, Barth C, Buchholz M, Lerch MM, Rolke M, Holzmann KH, Menke A, Fensterer H, Giehl K, Lohr M, Leder G, Iwamura T, Adler G, Gress TM (2003) Claudin-4 expression decreases invasiveness and metastatic potential of pancreatic cancer. Cancer Res 63(19):6265–6271

Miller CR, Buchsbaum DJ, Reynolds PN, Douglas JT, Gillespie GY, Mayo MS, Raben D, Curiel DT (1998) Differential susceptibility of primary and established human glioma cells to adenovirus infection: targeting via the epidermal growth factor receptor achieves fiber receptor-independent gene transfer. Cancer Res 58(24):5738–5748

Miwa N, Furuse M, Tsukita S, Niikawa N, Nakamura Y, Furukawa Y (2001) Involvement of claudin-1 in the beta-catenin/Tcf signaling pathway and its frequent upregulation in human colorectal cancers. Oncol Res 12(11–12):469–476

Mönkemüller K, Wex T, Kuester D, Fry LC, Kandulski A, Kropf S, Roessner A, Malfertheiner P (2012) Role of tight junction proteins in gastroesophageal reflux disease. BMC Gastroenterol 12 (1):128

Mori K, Iida S, Yamane-Ohnuki N, Kanda Y, Kuni-Kamochi R, Nakano R, Imai-Nishiya H, Okazaki A, Shinkawa T, Natsume A, Niwa R, Shitara K, Satoh M (2007) Non-fucosylated therapeutic antibodies: the next generation of therapeutic antibodies. Cytotechnology 55(2–3):109–114

Morin PJ (2005) Claudin proteins in human cancer: promising new targets for diagnosis and therapy. Cancer Res 65(21):9603–9606

Morton PE, Hicks A, Ortiz-Zapater E, Raghavan S, Pike R, Noble A, Woodfin A, Jenkins G, Rayner E, Santis G, Parsons M (2016) TNFα promotes CAR-dependent migration of leukocytes across epithelial monolayers. Sci Rep 6:26321

Murata M, Kojima T, Yamamoto T, Go M, Takano K, Osanai M, Chiba H, Sawada N (2005) Down-regulation of survival signaling through MAPK and Akt in occludin-deficient mouse hepatocytes in vitro. Exp Cell Res 310(1):140–151

Muto S, Hata M, Taniguchi J, Tsuruoka S, Moriwaki K, Saitou M, Furuse K, Sasaki H, Fujimura A, Imai M, Kusano E, Tsukita S, Furuse M (2010) Claudin-2-deficient mice are defective in the leaky and cation-selective paracellular permeability properties of renal proximal tubules. Proc Natl Acad Sci U S A 107(17):8011–8016

Naik MU, Naik TU, Suckow AT, Duncan MK, Naik UP (2008) Attenuation of junctional adhesion molecule-A is a contributing factor for breast cancer cell invasion. Cancer Res 68(7):2194–2203

Nakatsu G, Li X, Zhou H, Sheng J, Wong SH, Wu WKK, Ng SC, Tsoi H, Dong Y, Zhang N, He Y, Kang Q, Cao L, Wang K, Zhang J, Liang Q, Yu J, Sung JJY (2015) Gut mucosal microbiome across stages of colorectal carcinogenesis. Nat Commun 6(1):8727

Németh Z, Szász AM, Tátrai P, Németh J, Gyorffy H, Somorácz A, Szíjártó A, Kupcsulik P, Kiss A, Schaff Z (2009) Claudin-1, -2, -3, -4, -7, -8, and -10 protein expression in biliary tract cancers. J Histochem Cytochem 57(2):113–121

Nishizuka Y (1992) Intracellular signaling by hydrolysis of phospholipids and activation of protein kinase C. Science 258(5082):607–614

Nubel T, Preobraschenski J, Tuncay H, Weiss T, Kuhn S, Ladwein M, Langbein L, Zoller M (2009) Claudin-7 regulates EpCAM-mediated functions in tumor progression. Mol Cancer Res 7 (3):285–299

Okegawa T, Li Y, Pong RC, Bergelson JM, Zhou J, Hsieh JT (2000) The dual impact of coxsackie and adenovirus receptor expression on human prostate cancer gene therapy. Cancer Res 60 (18):5031–5036

Oliveira SS, Morgado-Díaz JA (2007) Claudins: multifunctional players in epithelial tight junctions and their role in cancer. Cell Mol Life Sci 64(1):17–28

Ono Y, Hiratsuka Y, Murata M, Takasawa A, Fukuda R, Nojima M, Tanaka S, Osanai M, Hirata K, Sawada N (2016) Claudins-4 and -7 might be valuable markers to distinguish hepatocellular carcinoma from cholangiocarcinoma. Virchows Arch 469(4):417–426

Osanai M, Murata M, Nishikiori N, Chiba H, Kojima T, Sawada N (2007) Occludin-mediated premature senescence is a fail-safe mechanism against tumorigenesis in breast carcinoma cells. Cancer Sci 98(7):1027–1034

Oshima T, Miwa H, Joh T (2008) Changes in the expression of claudins in active ulcerative colitis. J Gastroenterol Hepatol 23(Suppl 2):S146–S150

Oshima T, Koseki J, Chen X, Matsumoto T, Miwa H (2012) Acid modulates the squamous epithelial barrier function by modulating the localization of claudins in the superficial layers. Lab Investig 92(1):22–31

Osunkoya AO, Cohen C, Lawson D, Picken MM, Amin MB, Young AN (2009) Claudin-7 and claudin-8: immunohistochemical markers for the differential diagnosis of chromophobe renal cell carcinoma and renal oncocytoma. Hum Pathol 40(2):206–210

Paik J, Fierce Y, Treuting PM, Brabb T, Maggio-Price L (2013) High-fat diet-induced obesity exacerbates inflammatory bowel disease in genetically susceptible Mdr1a−/− male mice. J Nutr 143(8):1240–1247

Paine ER (2014) Colonoscopic evaluation in ulcerative colitis. Gastroenterol Rep (Oxf) 2(3):161–168

Pan XY, Wang B, Che YC, Weng ZP, Dai HY, Peng W (2007) Expression of claudin-3 and claudin-4 in normal, hyperplastic, and malignant endometrial tissue. Int J Gynecol Cancer 17 (1):233–241

Paoli P, Giannoni E, Chiarugi P (2013) Anoikis molecular pathways and its role in cancer progression. Biochim Biophys Acta 1833(12):3481–3498

Peerapen P, Thongboonkerd V (2013) p38 MAPK mediates calcium oxalate crystal-induced tight junction disruption in distal renal tubular epithelial cells. Sci Rep 3:1041

Peppercorn MA (2022) Clinical manifestations, diagnosis, and prognosis of Crohn disease in adults. In: Lamont TJ, Robson KM (eds) UpToDate

Philip R, Heiler S, Mu W, Büchler MW, Zöller M, Thuma F (2015) Claudin-7 promotes the epithelial-mesenchymal transition in human colorectal cancer. Oncotarget 6(4):2046–2063

Pope JL, Ahmad R, Bhat AA, Washington MK, Singh AB, Dhawan P (2014a) Claudin-1 overexpression in intestinal epithelial cells enhances susceptibility to adenamatous polyposis coli-mediated colon tumorigenesis. Mol Cancer 13(1):167

Pope JL, Bhat AA, Sharma A, Ahmad R, Krishnan M, Washington MK, Beauchamp RD, Singh AB, Dhawan P (2014b) Claudin-1 regulates intestinal epithelial homeostasis through the modulation of Notch-signalling. Gut 63(4):622–634

Prasad S, Mingrino R, Kaukinen K, Hayes KL, Powell RM, MacDonald TT, Collins JE (2005) Inflammatory processes have differential effects on claudins 2, 3 and 4 in colonic epithelial cells. Lab Investig 85(9):1139–1162

Principe DR, Timbers KE, Atia LG, Koch RM, Rana A (2021) TGFbeta signaling in the pancreatic tumor microenvironment. Cancers (Basel) 13(20)

Qi L, Shao W, Shi D (2013) JAM-2 siRNA intracellular delivery and real-time imaging by proton-sponge coated quantum dots. J Mater Chem B 1(5):654–660

Qi L, Liu J, Zhu H, Li Z, Lu K, Li T, Shi D (2014) Inhibition of glioma proliferation and migration by magnetic nanoparticle mediated JAM-2 silencing. J Mater Chem B 2(41):7168–7175
Qin M, Chen S, Yu T, Escuadro B, Sharma S, Batra RK (2003) Coxsackievirus adenovirus receptor expression predicts the efficiency of adenoviral gene transfer into non-small cell lung cancer xenografts. Clin Cancer Res 9(13):4992–4999
Qin Z, Fang D, Fang Y (2017) Low expression of occludin: a predictor of poor prognosis in esophageal squamous cell carcinoma. Int J Clin Exp Pathol 10(7):7451–7459
Ram AK, Pottakat B, Vairappan B (2018) Increased systemic zonula occludens 1 associated with inflammation and independent biomarker in patients with hepatocellular carcinoma. BMC Cancer 18(1):572
Rangel LB, Sherman-Baust CA, Wernyj RP, Schwartz DR, Cho KR, Morin PJ (2003) Characterization of novel human ovarian cancer-specific transcripts (HOSTs) identified by serial analysis of gene expression. Oncogene 22(46):7225–7232
Rao R (2009) Occludin phosphorylation in regulation of epithelial tight junctions. Ann N Y Acad Sci 1165:62–68
Rao RK, Samak G (2013) Bile duct epithelial tight junctions and barrier function. Tissue Barriers 1 (4):e25718
Rao RK, Basuroy S, Rao VU, Karnaky KJ Jr, Gupta A (2002) Tyrosine phosphorylation and dissociation of occludin-ZO-1 and E-cadherin-beta-catenin complexes from the cytoskeleton by oxidative stress. Biochem J 368(Pt 2):471–481
Reimer D, Steppan I, Wiedemair A, Concin N, Hofstetter G, Marth C, Müller-Holzner E, Zeimet AG (2007) Soluble isoforms but not the transmembrane form of coxsackie-adenovirus receptor are of clinical relevance in epithelial ovarian cancer. Int J Cancer 120(12):2568–2575
Rescigno M (2011) The intestinal epithelial barrier in the control of homeostasis and immunity. Trends Immunol 32(6):256–264
Resnick MB, Konkin T, Routhier J, Sabo E, Pricolo VE (2005) Claudin-1 is a strong prognostic indicator in stage II colonic cancer: a tissue microarray study. Mod Pathol 18(4):511–518
Rodilla V, Villanueva A, Obrador-Hevia A, Robert-Moreno A, Fernandez-Majada V, Grilli A, Lopez-Bigas N, Bellora N, Alba MM, Torres F, Dunach M, Sanjuan X, Gonzalez S, Gridley T, Capella G, Bigas A, Espinosa L (2009) Jagged1 is the pathological link between Wnt and Notch pathways in colorectal cancer. Proc Natl Acad Sci U S A 106(15):6315–6320
Rodriguez PL, Jiang S, Fu Y, Avraham S, Avraham HK (2014) The proinflammatory peptide substance P promotes blood-brain barrier breaching by breast cancer cells through changes in microvascular endothelial cell tight junctions. Int J Cancer 134(5):1034–1044
Roehlen N, Roca Suarez AA, El Saghire H, Saviano A, Schuster C, Lupberger J, Baumert TF (2020) Tight junction proteins and the biology of hepatobiliary disease. Int J Mol Sci 21(3)
Rosager AM, Sørensen MD, Dahlrot RH, Boldt HB, Hansen S, Lathia JD, Kristensen BW (2017) Expression and prognostic value of JAM-A in gliomas. J Neuro-Oncol 135(1):107–117
Rosenthal R, Milatz S, Krug SM, Oelrich B, Schulzke JD, Amasheh S, Günzel D, Fromm M (2010) Claudin-2, a component of the tight junction, forms a paracellular water channel. J Cell Sci 123 (Pt 11):1913–1921
Sahin U, Al-Batran S-E, Hozaeel W, Zvirbule Z, Freiberg-Richter J, Lordick F, Just M, Bitzer M, Thuss-Patience PC, Krilova A, Brass V, Mueller C, Daalman E, Tureci O (2015) IMAB362 plus zoledronic acid (ZA) and interleukin-2 (IL-2) in patients (pts) with advanced gastroesophageal cancer (GEC): Clinical activity and safety data from the PILOT phase I trial. J Clin Oncol 33 (15_suppl):e15079–e15079
Sahin U, Schuler M, Richly H, Bauer S, Krilova A, Dechow T, Jerling M, Utsch M, Rohde C, Dhaene K, Huber C, Tureci O (2018) A phase I dose-escalation study of IMAB362 (Zolbetuximab) in patients with advanced gastric and gastro-oesophageal junction cancer. Eur J Cancer 100:17–26
Sahin U, Tureci O, Manikhas G, Lordick F, Rusyn A, Vynnychenko I, Dudov A, Bazin I, Bondarenko I, Melichar B, Dhaene K, Wiechen K, Huber C, Maurus D, Arozullah A, Park JW, Schuler M, Al-Batran SE (2021) FAST: a randomised phase II study of zolbetuximab

(IMAB362) plus EOX versus EOX alone for first-line treatment of advanced CLDN18.2-positive gastric and gastro-oesophageal adenocarcinoma. Ann Oncol 32(5):609–619
Saito K, Sakaguchi M, Iioka H, Matsui M, Nakanishi H, Huh NH, Kondo E (2014) Coxsackie and adenovirus receptor is a critical regulator for the survival and growth of oral squamous carcinoma cells. Oncogene 33(10):1274–1286
Saitou M, Furuse M, Sasaki H, Schulzke JD, Fromm M, Takano H, Noda T, Tsukita S (2000) Complex phenotype of mice lacking occludin, a component of tight junction strands. Mol Biol Cell 11(12):4131–4142
Santin AD, Cane S, Bellone S, Palmieri M, Siegel ER, Thomas M, Roman JJ, Burnett A, Cannon MJ, Pecorelli S (2005) Treatment of chemotherapy-resistant human ovarian cancer xenografts in C.B-17/SCID mice by intraperitoneal administration of Clostridium perfringens enterotoxin. Cancer Res 65(10):4334–4342
Santin AD, Bellone S, Marizzoni M, Palmieri M, Siegel ER, McKenney JK, Hennings L, Comper F, Bandiera E, Pecorelli S (2007) Overexpression of claudin-3 and claudin-4 receptors in uterine serous papillary carcinoma: novel targets for a type-specific therapy using Clostridium perfringens enterotoxin (CPE). Cancer 109(7):1312–1322
Sarker MR, Carman RJ, McClane BA (1999) Inactivation of the gene (cpe) encoding Clostridium perfringens enterotoxin eliminates the ability of two cpe-positive C. perfringens type A human gastrointestinal disease isolates to affect rabbit ileal loops. Mol Microbiol 33(5):946–958
Satake S, Semba S, Matsuda Y, Usami Y, Chiba H, Sawada N, Kasuga M, Yokozaki H (2008) Cdx2 transcription factor regulates claudin-3 and claudin-4 expression during intestinal differentiation of gastric carcinoma. Pathol Int 58(3):156–163
Schaubeck M, Clavel T, Calasan J, Lagkouvardos I, Haange SB, Jehmlich N, Basic M, Dupont A, Hornef M, von Bergen M, Bleich A, Haller D (2016) Dysbiotic gut microbiota causes transmissible Crohn's disease-like ileitis independent of failure in antimicrobial defence. Gut 65(2):225–237
Schulzke JD, Gitter AH, Mankertz J, Spiegel S, Seidler U, Amasheh S, Saitou M, Tsukita S, Fromm M (2005) Epithelial transport and barrier function in occludin-deficient mice. Biochim Biophys Acta 1669(1):34–42
Seth A, Sheth P, Elias BC, Rao R (2007) Protein phosphatases 2A and 1 interact with occludin and negatively regulate the assembly of tight junctions in the CACO-2 cell monolayer. J Biol Chem 282(15):11487–11498
Sharma RK, Chheda ZS, Das Purkayastha BP, Gomez-Gutierrez JG, Jala VR, Haribabu B (2016) A spontaneous metastasis model reveals the significance of claudin-9 overexpression in lung cancer metastasis. Clin Exp Metastasis 33(3):263–275
Sharma A, Xu Y, Sung B, Vincent CT, Worgall T, Worgall S (2017) Regulation of the Coxsackie and adenovirus receptor expression is dependent on cystic fibrosis transmembrane regulator in airway epithelial cells. Cell Microbiol 19(2)
Sharma U, Murmu M, Barwal TS, Tuli HS, Jain M, Prakash H, Kaceli T, Jain A, Bishayee A (2022) A pleiotropic role of long non-coding RNAs in the modulation of Wnt/beta-catenin and PI3K/Akt/mTOR signaling pathways in esophageal squamous cell carcinoma: implication in chemotherapeutic drug response. Curr Oncol 29(4):2326–2349
Shayakhmetov DM, Li ZY, Ni S, Lieber A (2002) Targeting of adenovirus vectors to tumor cells does not enable efficient transduction of breast cancer metastases. Cancer Res 62(4):1063–1068
Sheehan GM, Kallakury BV, Sheehan CE, Fisher HA, Kaufman RP Jr, Ross JS (2007) Loss of claudins-1 and -7 and expression of claudins-3 and -4 correlate with prognostic variables in prostatic adenocarcinomas. Hum Pathol 38(4):564–569
Sheng S, Hu Y, Yu F, Tong W, Wang S, Cai Y, Zhu J (2020) circKIF4A sponges miR-127 to promote ovarian cancer progression. Aging (Albany NY) 12(18):17921–17929
Sibbald B (2001) Death but one unintended consequence of gene-therapy trial. CMAJ 164 (11):1612
Simon DB, Lu Y, Choate KA, Velazquez H, Al-Sabban E, Praga M, Casari G, Bettinelli A, Colussi G, Rodriguez-Soriano J, McCredie D, Milford D, Sanjad S, Lifton RP (1999) Paracellin-1, a

renal tight junction protein required for paracellular Mg2+ resorption. Science 285(5424):103–106

Singh AB, Sharma A, Dhawan P (2010) Claudin family of proteins and cancer: an overview. J Oncol 2010:541957

Singh AB, Sharma A, Dhawan P (2012) Claudin-1 expression confers resistance to anoikis in colon cancer cells in a Src-dependent manner. Carcinogenesis 33(12):2538–2547

Singh P, Toom S, Huang Y (2017) Anti-claudin 18.2 antibody as new targeted therapy for advanced gastric cancer. J Hematol Oncol 10(1):105

Soini Y, Tommola S, Helin H, Martikainen P (2006) Claudins 1, 3, 4 and 5 in gastric carcinoma, loss of claudin expression associates with the diffuse subtype. Virchows Arch 448(1):52–58

Suh Y, Yoon CH, Kim RK, Lim EJ, Oh YS, Hwang SG, An S, Yoon G, Gye MC, Yi JM, Kim MJ, Lee SJ (2013) Claudin-1 induces epithelial-mesenchymal transition through activation of the c-Abl-ERK signaling pathway in human liver cells. Oncogene 32(41):4873–4882

Suren D, Yildirim M, Kaya V, Elal R, Selcuk OT, Osma U, Yildiz M, Gunduz S, Sezer C (2015) Expression patterns of claudins 1, 4, and 7 and their prognostic significance in nasopharyngeal carcinoma. J BUON 20(1):212–217

Suzuki M, Kato-Nakano M, Kawamoto S, Furuya A, Abe Y, Misaka H, Kimoto N, Nakamura K, Ohta S, Ando H (2009) Therapeutic antitumor efficacy of monoclonal antibody against Claudin-4 for pancreatic and ovarian cancers. Cancer Sci 100(9):1623–1630

Tabariès S, Dupuy F, Dong Z, Monast A, Annis MG, Spicer J, Ferri LE, Omeroglu A, Basik M, Amir E, Clemons M, Siegel PM (2012) Claudin-2 promotes breast cancer liver metastasis by facilitating tumor cell interactions with hepatocytes. Mol Cell Biol 32(15):2979–2991

Tabariès S, Annis MG, Hsu BE, Tam CE, Savage P, Park M, Siegel PM (2015) Lyn modulates Claudin-2 expression and is a therapeutic target for breast cancer liver metastasis. Oncotarget 6(11):9476–9487

Takai E, Tan X, Tamori Y, Hirota M, Egami H, Ogawa M (2005) Correlation of translocation of tight junction protein Zonula occludens-1 and activation of epidermal growth factor receptor in the regulation of invasion of pancreatic cancer cells. Int J Oncol 27(3):645–651

Takala H, Saarnio J, Wiik H, Soini Y (2007) Claudins 1, 3, 4, 5 and 7 in esophageal cancer: loss of claudin 3 and 4 expression is associated with metastatic behavior. APMIS 115(7):838–847

Takehara M, Nishimura T, Mima S, Hoshino T, Mizushima T (2009) Effect of claudin expression on paracellular permeability, migration and invasion of colonic cancer cells. Biol Pharm Bull 32(5):825–831

Tamura A, Hayashi H, Imasato M, Yamazaki Y, Hagiwara A, Wada M, Noda T, Watanabe M, Suzuki Y, Tsukita S (2011) Loss of claudin-15, but not claudin-2, causes Na+ deficiency and glucose malabsorption in mouse small intestine. Gastroenterology 140(3):913–923

Tanaka A, Mertens JC (2016) Ulcerative colitis with and without primary sclerosing cholangitis: two different diseases? Inflamm Intest Dis 1(1):9–14

Tanaka M, Shibahara J, Fukushima N, Shinozaki A, Umeda M, Ishikawa S, Kokudo N, Fukayama M (2011) Claudin-18 is an early-stage marker of pancreatic carcinogenesis. J Histochem Cytochem 59(10):942–952

Tanaka A, Ishikawa S, Ushiku T, Yamazawa S, Katoh H, Hayashi A, Kunita A, Fukayama M (2018) Frequent CLDN18-ARHGAP fusion in highly metastatic diffuse-type gastric cancer with relatively early onset. Oncotarget 9(50):29336–29350

Thompson EM, Pishko GL, Muldoon LL, Neuwelt EA (2013) Inhibition of SUR1 decreases the vascular permeability of cerebral metastases. Neoplasia 15(5):535–543

Thompson EM, Halvorson K, McLendon R (2018) Sulfonylurea receptor 1 expression is variable in adult and pediatric brain tumors. Clin Neuropathol 37(5):221–227

Thuma F, Heiler S, Schnolzer M, Zoller M (2016) Palmitoylated claudin7 captured in glycolipid-enriched membrane microdomains promotes metastasis via associated transmembrane and cytosolic molecules. Oncotarget 7(21):30659–30677

Tokumasu R, Yamaga K, Yamazaki Y, Murota H, Suzuki K, Tamura A, Bando K, Furuta Y, Katayama I, Tsukita S (2016) Dose-dependent role of claudin-1 in vivo in orchestrating features of atopic dermatitis. Proc Natl Acad Sci U S A 113(28):E4061–E4068

Torres-Martínez AC, Gallardo-Vera JF, Lara-Holguin AN, Montaño LF, Rendón-Huerta EP (2017) Claudin-6 enhances cell invasiveness through claudin-1 in AGS human adenocarcinoma gastric cancer cells. Exp Cell Res 350(1):226–235

Tsai PY, Zhang B, He WQ, Zha JM, Odenwald MA, Singh G, Tamura A, Shen L, Sailer A, Yeruva S, Kuo WT, Fu YX, Tsukita S, Turner JR (2017) IL-22 upregulates epithelial claudin-2 to drive diarrhea and enteric pathogen clearance. Cell Host Microbe 21(6):671–681. e674

Tureci O, Sahin U, Schulze-Bergkamen H, Zvirbule Z, Lordick F, Koeberle D, Thuss-Patience P, Ettrich T, Arnold D, Bassermann F, Al-Batran SE, Wiechen K, Dhaene K, Maurus D, Gold M, Huber C, Krivoshik A, Arozullah A, Park JW, Schuler M (2019) A multicentre, phase IIa study of zolbetuximab as a single agent in patients with recurrent or refractory advanced adenocarcinoma of the stomach or lower oesophagus: the MONO study. Ann Oncol 30(9):1487–1495

Tzelepi VN, Tsamandas AC, Vlotinou HD, Vagianos CE, Scopa CD (2008) Tight junctions in thyroid carcinogenesis: diverse expression of claudin-1, claudin-4, claudin-7 and occludin in thyroid neoplasms. Mod Pathol 21(1):22–30

Ungaro R, Mehandru S, Allen PB, Peyrin-Biroulet L, Colombel JF (2017) Ulcerative colitis. Lancet 389(10080):1756–1770

Utech M, Ivanov AI, Samarin SN, Bruewer M, Turner JR, Mrsny RJ, Parkos CA, Nusrat A (2005) Mechanism of IFN-gamma-induced endocytosis of tight junction proteins: myosin II-dependent vacuolarization of the apical plasma membrane. Mol Biol Cell 16(10):5040–5052

van de Wetering M, Sancho E, Verweij C, de Lau W, Oving I, Hurlstone A, van der Horn K, Batlle E, Coudreuse D, Haramis AP, Tjon-Pon-Fong M, Moerer P, van den Born M, Soete G, Pals S, Eilers M, Medema R, Clevers H (2002) The beta-catenin/TCF-4 complex imposes a crypt progenitor phenotype on colorectal cancer cells. Cell 111(2):241–250

Van Itallie C, Rahner C, Anderson JM (2001) Regulated expression of claudin-4 decreases paracellular conductance through a selective decrease in sodium permeability. J Clin Invest 107(10):1319–1327

Vindrieux D, Le Corre L, Hsieh JT, Métivier R, Escobar P, Caicedo A, Brigitte M, Lazennec G (2011) Coxsackie and adenovirus receptor is a target and a mediator of estrogen action in breast cancer. Endocr Relat Cancer 18(3):311–321

Vorburger SA, Hunt KK (2002) Adenoviral gene therapy. Oncologist 7(1):46–59

Wang H, Yang X (2015) The expression patterns of tight junction protein claudin-1, −3, and −4 in human gastric neoplasms and adjacent non-neoplastic tissues. Int J Clin Exp Pathol 8(1):881–887

Wang Y, Wang S, Bao Y, Ni C, Guan N, Zhao J, Salford LG, Widegren B, Fan X (2006) Coxsackievirus and adenovirus receptor expression in non-malignant lung tissues and clinical lung cancers. J Mol Histol 37(3–4):153–160

Wang X, Tully O, Ngo B, Zitin M, Mullin JM (2011) Epithelial tight junctional changes in colorectal cancer tissues. Sci World J 11:826–841

Wang K, Xu C, Li W, Ding L (2018) Emerging clinical significance of claudin-7 in colorectal cancer: a review. Cancer Manag Res 10:3741–3752

Weber CR, Nalle SC, Tretiakova M, Rubin DT, Turner JR (2008) Claudin-1 and claudin-2 expression is elevated in inflammatory bowel disease and may contribute to early neoplastic transformation. Lab Investig 88(10):1110–1120

Weiner LM, Surana R, Wang S (2010) Monoclonal antibodies: versatile platforms for cancer immunotherapy. Nat Rev Immunol 10(5):317–327

Xia P, Wang W, Bai Y (2014) Claudin-7 suppresses the cytotoxicity of TRAIL-expressing mesenchymal stem cells in H460 human non-small cell lung cancer cells. Apoptosis 19 (3):491–505

Xu XR, Liu CQ, Feng BS, Liu ZJ (2014) Dysregulation of mucosal immune response in pathogenesis of inflammatory bowel disease. World J Gastroenterol 20(12):3255–3264

Xu K, Sun G, Li M, Chen H, Zhang Z, Qian X, Li P, Xu L, Huang W, Wang X (2019) Glibenclamide targets sulfonylurea receptor 1 to inhibit p70S6K activity and upregulate

KLF4 expression to suppress non-small cell lung carcinoma. Mol Cancer Ther 18(11):2085–2096

Yamaguchi H, Kojima T, Ito T, Kimura Y, Imamura M, Son S, Koizumi J, Murata M, Nagayama M, Nobuoka T, Tanaka S, Hirata K, Sawada N (2010) Transcriptional control of tight junction proteins via a protein kinase C signal pathway in human telomerase reverse transcriptase-transfected human pancreatic duct epithelial cells. Am J Pathol 177(2):698–712

Yang J, Liu X, Yuan X, Wang Z (2015) miR-99b promotes metastasis of hepatocellular carcinoma through inhibition of claudin 11 expression and may serve as a prognostic marker. Oncol Rep 34 (3):1415–1423

Yang P, Zhang M, Liu X, Pu H (2017) MicroRNA-421 promotes the proliferation and metastasis of gastric cancer cells by targeting claudin-11. Exp Ther Med 14(3):2625–2632

Youakim A, Ahdieh M (1999) Interferon-gamma decreases barrier function in T84 cells by reducing ZO-1 levels and disrupting apical actin. Am J Phys 276(5):G1279–G1288

Yu LC-H (2018) Microbiota dysbiosis and barrier dysfunction in inflammatory bowel disease and colorectal cancers: exploring a common ground hypothesis. J Biomed Sci 25(1):79

Yu S, Zhang Y, Li Q, Zhang Z, Zhao G, Xu J (2019) CLDN6 promotes tumor progression through the YAP1-snail1 axis in gastric cancer. Cell Death Dis 10(12):949

Zeisel MB, Crouchet E, Baumert TF, Schuster C (2015) Host-targeting agents to prevent and cure hepatitis C virus infection. Viruses 7(11):5659–5685

Zeisel MB, Dhawan P, Baumert TF (2019) Tight junction proteins in gastrointestinal and liver disease. Gut 68(3):547–561

Zeissig S, Bürgel N, Günzel D, Richter J, Mankertz J, Wahnschaffe U, Kroesen AJ, Zeitz M, Fromm M, Schulzke JD (2007) Changes in expression and distribution of claudin 2, 5 and 8 lead to discontinuous tight junctions and barrier dysfunction in active Crohn's disease. Gut 56(1):61–72

Zeng MY, Inohara N, Nuñez G (2017) Mechanisms of inflammation-driven bacterial dysbiosis in the gut. Mucosal Immunol 10(1):18–26

Zhang M, Luo W, Huang B, Liu Z, Sun L, Zhang Q, Qiu X, Xu K, Wang E (2013) Overexpression of JAM-A in non-small cell lung cancer correlates with tumor progression. PLoS One 8(11): e79173

Zhang SJ, Feng JF, Wang L, Guo W, Du YW, Ming L, Zhao GQ (2014) miR-1303 targets claudin-18 gene to modulate proliferation and invasion of gastric cancer cells. Dig Dis Sci 59(8):1754–1763

Zhang Y, Xia F, Liu X, Yu Z, Xie L, Liu L, Chen C, Jiang H, Hao X, He X, Zhang F, Gu H, Zhu J, Bai H, Zhang CC, Chen GQ, Zheng J (2018) JAM3 maintains leukemia-initiating cell self-renewal through LRP5/AKT/β-catenin/CCND1 signaling. J Clin Invest 128(5):1737–1751

Zhang J, Dong R, Shen L (2020) Evaluation and reflection on claudin 18.2 targeting therapy in advanced gastric cancer. Chin J Cancer Res 32(2):263–270

Zhao C, Lu F, Chen H, Zhao X, Sun J, Chen H (2014) Dysregulation of JAM-A plays an important role in human tumor progression. Int J Clin Exp Pathol 7(10):7242–7248

Zihni C, Mills C, Matter K, Balda MS (2016) Tight junctions: from simple barriers to multifunctional molecular gates. Nat Rev Mol Cell Biol 17(9):564–580

Zuo L, Kuo WT, Turner JR (2020) Tight junctions as targets and effectors of mucosal immune homeostasis. Cell Mol Gastroenterol Hepatol 10(2):327–340

Zwiers A, Fuss IJ, Leijen S, Mulder CJ, Kraal G, Bouma G (2008) Increased expression of the tight junction molecule claudin-18 A1 in both experimental colitis and ulcerative colitis. Inflamm Bowel Dis 14(12):1652–1659

2 Regulation of Tight Junction by Cadherin Adhesion and Its Implication in Inflammation and Cancer

S. M. Nasir Uddin, Asfia Sultana, Asma Fatima, Anupriya M. Geethakumari, and Kabir H. Biswas

Abstract

Epithelial cells interact with their cellular microenvironment through the formation of different interactions between the plasma membrane-localized receptors of adjacent cells and the extracellular matrix. Of these interactions, tight junctions (TJs), also known as zonula occludens, are the semipermeable intercellular adhesion complexes that control both epithelial and endothelial paracellular permeability preventing the intermixing of apical and basolateral membrane components and, as a result, give rise to morphologically and functionally distinct apical and basolateral domains, leading to epithelial cell polarization. TJ-associated transmembrane proteins, such as occludin, claudins, and junctional adhesion molecules (JAM), encompass an extracellular domain that forms distinct branched strands at the core of the tight junction structure, whereas peripheral membrane protein, such as zonula occludens-1 (ZO-1), interacts with the transmembrane proteins and aids in their anchoring to the cytoskeleton. Functionally, TJs are involved in bidirectional signaling between adhering cells that regulate cellular differentiation, migration, proliferation, and survival enabling them to play a vital role in a number of physiological processes such as innate immunity, aside from the barrier function in the epithelial tissue and mutations in TJ proteins can result in different diseases including cancers, infections, and allergies. Interestingly, TJs have been found to interact with cadherin-based adherens junctions (AJs) and desmosomes that can fine-tune their assembly as well as signaling activities. In this chapter, we discuss the structure and function

S. M. Nasir Uddin, Asfia Sultana and Asma Fatima contributed equally with all other contributors.

S. M. Nasir Uddin · A. Sultana · A. Fatima · A. M. Geethakumari · K. H. Biswas (✉)
Division of Biological and Biomedical Sciences, College of Health & Life Sciences, Hamad Bin Khalifa University, Doha, Qatar
e-mail: kbiswas@hbku.edu.qa

A. A. Bhat et al. (eds.), *Tight Junctions in Inflammation and Cancer*,
https://doi.org/10.1007/978-981-99-2415-8_2

of both TJs and AJs, modes of their interaction, and their role in inflammation and cancer.

Keywords

Actomyosin · Adhesion · Cadherin · Cytoskeleton · Tight junctions

2.1 Introduction

The three-dimensional epithelial tissue organization is attributed to the cellular adhesion machinery consisting of various types of cell adhesion complexes, including TJs, AJs, gap junctions, and desmosomes (Biswas and Zaidel-Bar 2017). These specialized cell adhesion complexes facilitate and regulate the architecture of the cells, including their shape, structure, and microenvironment, through cellular adhesion and barrier formation (Ikenouchi et al. 2007; Hartsock and Nelson 2008; Rasool et al. 2021; Biswas and Groves 2016). Specifically, TJs are multifunctional intercellular assemblies bordering the apical and basolateral regions of epithelial cell membrane that act as barriers regulating cell permeability and maintaining cell polarity by preventing lateral diffusion of apical and basolateral membrane components. TJs are complexes of integral transmembrane proteins such as occludin, claudin, tricellulin, and junctional adhesion molecules (JAM) (Fig. 2.1). Zonula occludens (ZO)—a peripheral membrane protein—connects the actin cytoskeleton with the transmembrane proteins (Mège and Ishiyama 2017). These proteins are multifunctional and are associated with various cellular signaling interactions. Occludin and claudin transduce cellular signaling that regulate cell differentiation, growth, proliferation, and polarity (Shin et al. 2006). Despite continued research, the signaling mechanism regulating TJs still remains unclear due to the complexity and versatility in their structure and function (Martin and Jiang 2009; Soini 2011).

Of all the associated proteins, occludin is the first identified TJ transmembrane protein with four transmembrane domains expressed in the epithelial cells. The first extracellular loop is associated with paracellular permeability, while the second extracellular loop localizes occludin to the TJ. Occludin dimerization is controlled by the "C" terminal while the interaction of occludin with ZO-1 mediates its intracellular trafficking to the plasma membrane. Previous studies have shown that binding of occludin with claudin is required for stabilizing the TJ strands. However, knockdown and knockout of occludin have been shown to result in differential TJ formation (Cummins 2012).

Claudins (CLDN) are small transmembrane proteins categorized as classic claudins (1–10, 14, 15, 17, 19) and nonclassic claudins (11–13, 16, 18, 20–24), consisting of two extracellular loops and the PDZ domain, which helps in binding to different scaffolding proteins. The paracellular tightness and selective ion

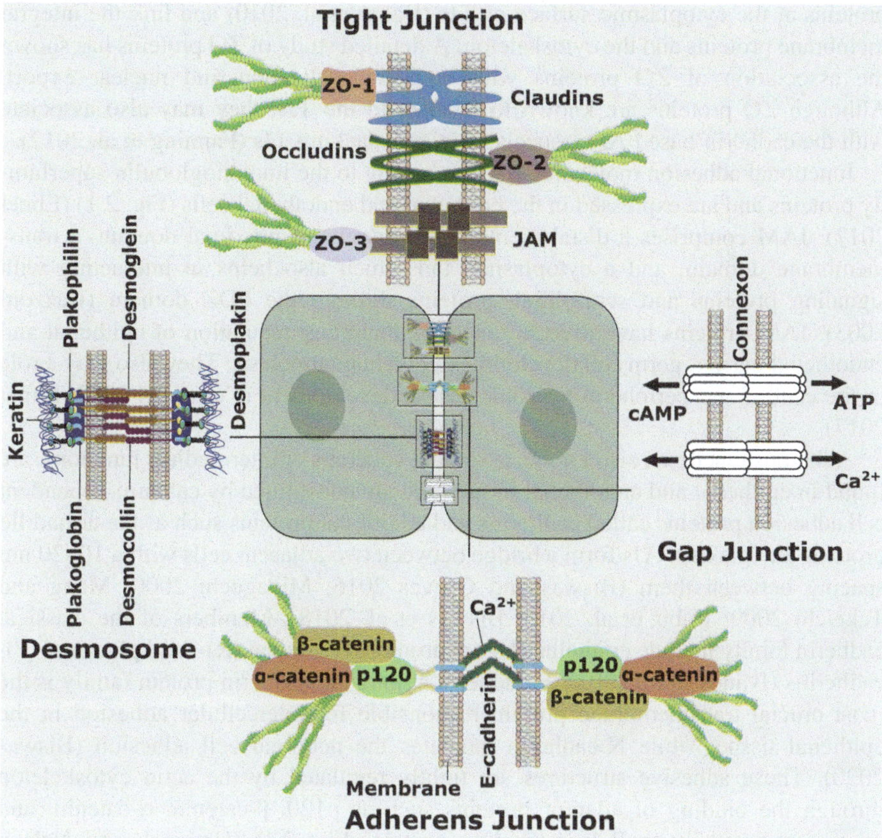

Fig. 2.1 Various intercellular interactions formed by epithelial cells. Upper panel: Tight junctions at the epithelial cell membranes represent TJ proteins—claudins, occludin, and JAM. ZO proteins connect the actin filaments with the TJ proteins. Right panel: Gap junction—intercellular junction connecting the cytoplasm of two cells. Connexin-hexamer protein in gap junction. Left panel: Desmosomes—stronger cell-cell adhesion type on the lateral sides of plasma membrane consists of desmoglein, desmocollin along with plakoglobin and plakophilin mediatory proteins. Lower panel: Cell-cell adhesion mediated by E-cadherin connected to the actin cytoskeleton through α- and β-catenin.

permeability are mediated by the first extracellular loop while the second extracellular loop narrows the paracellular cleft holding the opposing cell membrane (Fig. 2.1). CLDN3 and CLDN4 are important among other claudin proteins as they help in maintaining motility, invasion, and anoikis resistance. These transmembrane proteins are crucial for maintaining physiological homeostasis (Ivanov et al. 2001; Krause et al. 2008).

Zonula occludens (ZO proteins) are the scaffolding proteins belonging to the membrane-associated guanylate kinase (MAGUK) family proteins (Fig. 2.1) (González-Mariscal et al. 2000). They provide the base for the assembly of various proteins at the cytoplasmic surface of TJs (Bauer et al. 2010) and link the integral membrane proteins and the cytoskeleton. A detailed study of ZO proteins has shown the association of ZO proteins with nuclear localization and nuclear export. Although ZO proteins are known to localize to the TJs, they may also associate with the cadherin-based AJs, especially in cells lacking TJs (Fanning et al. 2012).

Junctional adhesion molecules (JAMs) belong to the immunoglobulin superfamily proteins and are expressed in the epithelial and endothelial cells (Fig. 2.1) (Ebnet 2017). JAM comprises a distal membrane domain and proximal domain, a transmembrane domain, and a cytoplasmic tail which also helps in interacting with signaling proteins and scaffolding proteins through the PDZ domain (Bazzoni 2003). JAM proteins have diverse functions including regulation of epithelial and endothelial barrier, germ cell development, and hematopoiesis. They also play a role in the central and peripheral nervous system development (Cummins 2012; Ebnet 2017).

Adherens junctions, also known as zonula adherens or intermediate junctions, are found in epithelial and endothelial tissues and are constituted by calcium-dependent cell adhesion proteins called cadherins and additional proteins such as the armadillo proteins and plakins. AJs form a bridge between two adjacent cells with a 10–20 nm spacing between them (Biswas and Groves 2016; Miyaguchi 2000; Meng and Takeichi 2009; Kabir et al. 2018; Biswas et al. 2018). Members of the classical cadherin family include epithelial (E)-, neuronal (N)-, retinal (R)-, and placental (P)-cadherins (Ivanov et al. 2001). E-cadherin among the cadherin protein family is the most crucial transmembrane protein responsible for intercellular adhesion in the epithelial tissue, while N-cadherin mediates the neuronal cell adhesion (Biswas 2020). These adhesive structures are tightly regulated by the actin cytoskeleton through the binding of adaptor proteins such as p120, β-catenin, α-catenin, and other associated intracellular signaling protein (Fig. 2.1) (Hartsock and Nelson 2008). Any alteration or interruptions in the communication between cadherins and the actin cytoskeleton or changes in cell membrane dynamics are known to affect cadherin organization on the cell membrane, thus altering the cell-cell adhesion (Biswas 2020). Nonclassical cadherins, including desmosomal cadherins, protocadherin, and cadherin-related molecules, play an essential role in cell polarity establishment and functioning (Harris and Tepass 2010; Altamash et al. 2021). The interaction of the cadherin-catenin complex ensures cellular integrity and plays a vital role in organ development and tissue homeostasis (Biswas 2020; Chen et al. 2018, 2021). Any change in the mechanism or dysregulation of this interconnected biological system can potentially lead to tumor initiation and metastasis.

The TJs and AJs are highly interdependent in that the formation of TJs rely on AJs while mutations in TJ proteins affect the maturation of AJs (Ikenouchi et al. 2007). However, the mechanism of this mutual regulation is yet to be completely understood. This chapter focuses on the interaction of the TJs and AJs and their role in the reversible biological phenomenon—epithelial-mesenchymal transition (EMT) and the implications of their dysregulation on the cellular microenvironment changes leading to diseases such as cancer or a response to infection or physical injury. The chapter also covers the role of TJs and the associated proteins in the manifestation of different types of cancers.

2.2 Interaction Between TJs and AJs

The cell-cell adhesion and barrier formation in the epithelial and other tissues are dependent on the AJs and TJs (Fig. 2.2). The epithelial tissue monolayer formation is initiated by the actin cytoskeleton-mediated protrusion of cells that reach to the surrounding cells, followed by homophilic interaction of E-cadherins, leading to the assembly of AJs (Biswas and Zaidel-Bar 2017; Rasool et al. 2021; Biswas 2020; Biswas et al. 2015a, b, 2016). Following this, TJs start appearing on one side of the AJ resulting in the development of polarity in the cells. The assembly and maintenance of these two types of junctions are highly interlinked and dependent on each other (Twiss et al. 2012). Specifically, one of the key aspects of E-cadherin-based adhesions, i.e., mechanosensing, is thought to be involved in the regulation of TJ formation as binding of vinculin (a mechanosensitive protein localized to E-cadherin-based adhesion) to the E-cadherin-based adhesion complex affects the rate of barrier formation in epithelial membranes (Twiss et al. 2012). For example, in cardiomyocytes, vinculin localizes to the intercalated discs, which are regions of adhesion junctions, helping in binding of the actin cytoskeleton to the sarcolemma. At intercalated discs, vinculin colocalizes with the TJ protein, ZO-1, and their direct interaction is evident from yeast two-hybrid studies (Twiss et al. 2012). Moreover, deletion of the vinculin gene reduces the expression levels of ZO-1, Cx43, talin, and

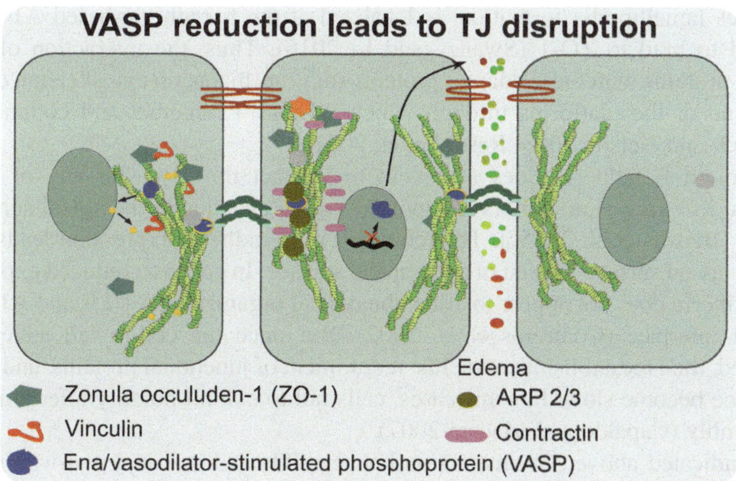

Fig. 2.2 Intracellular AJ-TJ interaction through actin cytoskeleton. Schematic representation of the interaction between TJ and cadherin-based junctions through intracellular proteins. Interaction between TJ and cadherin-based AJs through the actin cytoskeleton. The vinculin localizes at the cadherin-based AJs; it binds the α-catenin to the actin cytoskeleton, while in TJ assembly, it interacts with ZO-1 protein. Another actin-binding protein, VASP, is involved in the actin polymerization at the cadherin-based junction. But VASP knockout results in the disruption of the barrier between two cells. Cortactin interacts with both TJ proteins and cadherin-based AJ protein complex

β1D-integrin, an integrin isoform which is specifically expressed in muscles, proteins. Additionally, loss of vinculin is reported to suppress PTEN signalling, which affects cell polarity and paracellular permeability (Zemljic-Harpf et al. 2014). In addition, the cytoplasmic scaffold proteins of both AJs and TJs are observed to interact directly. For example, the TJ scaffold protein, ZO-1, is known to bind to alpha-catenin at its C-terminus, and this interaction is believed to be required for TJ assembly (Maiers et al. 2013). Indeed, inhibition of this interaction results in the reduction of the junctional resistance, disruption in the localization of occludin and ZO-1 at the cell-cell junction although cadherin-based AJ assembly is not affected (Maiers et al. 2013). Further, with the help of immunofluorescence and immunoelectron microscopy, it has been suggested that alpha-catenin is involved in the mobilization of ZO-1 from cytoplasm to the periphery of the cell (Maiers et al. 2013).

Besides the crosstalk during their assembly, these two adhesive structures are physically linked through the cortical actin cytoskeleton. The actin-binding protein Ena/vasodilator-stimulated phosphoprotein (VASP) that promotes actin filament polymerization is highly localized at the cadherin-based AJs and is involved in regulating actin cytoskeleton polymerization at the AJs (Scott et al. 2006). The absence of the VASP (Ena/VASP knock out) can result in diseases such as edema, hemorrhage, and embryonic lethality, apparently due to its impact in the formation and maintenance of TJs (Furman et al. 2007). Similarly, cortactin, a protein that promotes cell motility by recruiting the ARP2/3 complex to the actin filament and facilitates lamellipodia formation, is localized at the E-cadherin-based AJs and is reported to bind to ZO-1 (Swaney and Li 2016). Thus, the interaction of actin-binding proteins with these adaptor proteins suggests that actin cytoskeleton dynamics is one of the platforms through which AJ and TJ interact and communicate (Hartsock and Nelson 2008; Rasool et al. 2021).

Although E-cadherin does not seem to interact directly with any of the TJ proteins, its expression confers the morphological changes in the apical surface of the cell (Biswas et al. 2015c). The reduction in E-cadherin expression leads to the swelled apical surface instead of a flat apical surface. In contrast, reduced expression of E-cadherin does not appear to affect the typical organization of TJs and AJs at the cell-cell interface (Contreras et al. 2002). But once the cell-to-cell adhesion is disrupted, then reestablishment of TJs, recruitment of junctional proteins, and barrier resistance become slow. In some cases, cells fail to exhibit polarity after junctional disassembly (Capaldo and Macara 2007).

As indicated above, the interaction between TJs and AJs is a two-way process (Fig. 2.3). TJ proteins control the cell-cell adhesion by affecting AJ protein expression in the cells (Hartsock and Nelson 2008). The claudin-7, a TJ transmembrane protein, supports the assembly of AJs by positively regulating the expression of E-cadherin (Lioni et al. 2007) while claudin-1 regulates the expression of E-cadherin as dysregulation of claudin1 disrupts the E-cadherin expression by upregulating repressor ZEB-1 (Singh et al. 2011). AJs also regulate the formation of TJs by controlling EGFR activity and the lipid content of cell membranes (Rübsam et al. 2017; Shigetomi et al. 2018). For instance, AJs increase the cholesterol level in the

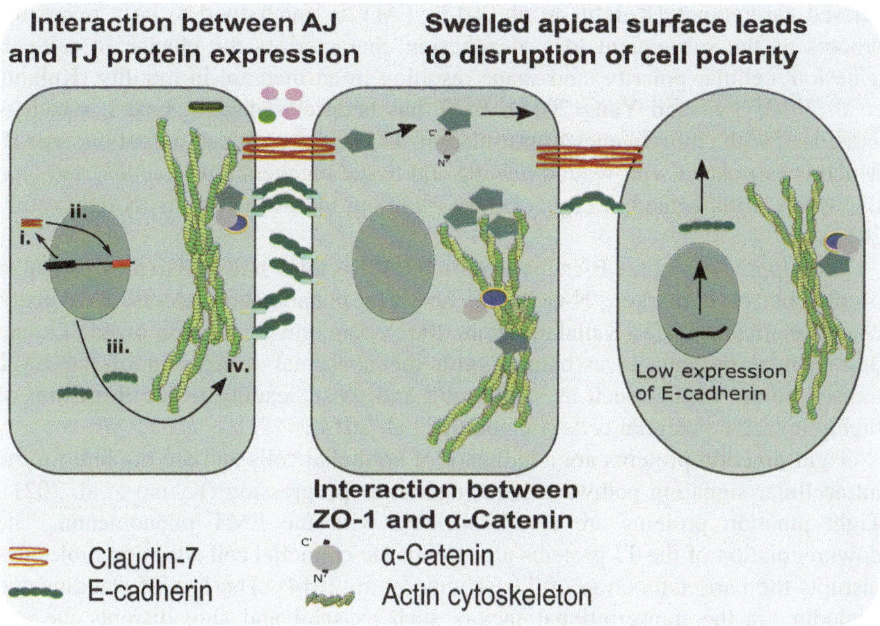

Fig. 2.3 Interaction between TJs and AJs. The protein involved in the assembly of TJ and cadherin-based AJ, affecting the activity of each other through direct binding. The expression of TJ proteins and cadherin-based AJ proteins are interlinked in the following way: (i) Low expression of claudin-7, a TJ transmembrane protein; (ii) affecting the expression of E-cadherin and (iii) thus, reduction in E-cadherin expression; (iv) that result in decrease in number of E-cadherin-based junction. A TJ intracellular protein, ZO-1, binds to α-catenin (protein present in the cadherin-based adhesion complex) at its C-terminus, and this binding interaction is required for the TJ assembly. Low E-cadherin expression leads to the swelled apical surface instead of a flat apical surface

cell membranes, thus, maintaining the stability of claudins on the cell membrane since a decrease in membrane cholesterol levels is known to increase the rate of claudin endocytosis (Shigetomi et al. 2018).

2.3 Epithelial to Mesenchymal Transition (EMT)

AJ is one of the key factors that regulate cell migration and maintaining epithelial homeostasis (Ozawa and Kobayashi 2014). The AJ-mediated cellular adhesion is maintained by linking of the transmembrane protein to the actin cytoskeleton. Further, a loss of E-cadherin expression or mutations affecting the function of the protein is attributed to epithelial-mesenchymal transition (EMT) and tumor cell invasion (Capaldo and Macara 2007; Ozawa and Kobayashi 2014). EMT is considered to be a normal biological and developmental process that is involved primarily in processes such as embryogenesis and wound healing as well as in epithelium-derived carcinomas (Knights et al. 2012). EMT is understood to be a reversible process as the subsequent loss of adhesion characterizes the change in cell-cell adhesion, cellular polarity, and shape resulting in an increase in motility (Knights et al. 2012; Tsai and Yang 2013). EMT has been classified as type I, which is associated with embryogenesis, gastrulation, and neural crest cell migration, type II, which is associated with wound healing and tissue regeneration in adults, and type III, which is implicated in cancer progression and tumor metastasis (Knights et al. 2012).

The suppression of the E-cadherin during EMT is known to lead to the expression of mesenchymal marker, N-cadherin, and this phenomenon serves as a major characteristics of EMT (Kallakury et al. 2001). The downregulation of AJ, TJ, and desmosomal proteins is associated with the abnormal expression of the EMT transcriptional factors such as snail, snug and twist, leading to the formation of highly invasive potential cells (Lamouille et al. 2014).

Tight junction proteins are a hallmark of epithelial cells and are the hub for the intracellular signaling pathways inducing cancer progression (Kyuno et al. 2021). Tight junction proteins are also associated with the EMT phenomenon. The downregulation of the TJ proteins along with the epithelial cell adhesion molecules disrupts the barrier function of TJs (Kyuno et al. 2014). The loss of claudins and occludin via the transcriptional factors such as snail and slug disrupts the cell polarity and the junctions causing loss of barrier function that leads to poor selective permeability (Kyuno et al. 2014; Quail and Joyce 2013). These transcription factors are known to bind to the claudin motifs and negatively regulate their function in many cancers. The pancreatic cancer is one such example that shows a reduced expression of the claudin-1 protein (Hotz et al. 2007). EMT is also associated with the repression of the genes encoding claudin, occludin, and E-cadherin leading to metastasis and cancer (Boyer et al. 2000).

Different types of cancer cells show differential expression of the E-cadherin and N-cadherin genes. In this regard, the methylation of the E-cadherin gene promoter in breast cancer cells is understood to lead to the transition from E-cadherin expression to N-cadherin expression, resulting in the loss of E-cadherin's tumor suppressor activity (Knights et al. 2012; Davis et al. 2003). One of the primary reasons for the downregulation of the adhesion genes in EMT is active transcriptional repression. Many carcinomas have shown evidence of the collective repression of the E-cadherin, α-catenin, β-catenin, and p120-catenin (Davis et al. 2003). One of the examples explaining the above is adenocarcinoma. This coordinated repression mechanism and the downregulation or overexpression of the AJ components involved in signaling pathways may produce different outcomes associated with cancer invasion and metastasis (Kallakury et al. 2001).

2.4 Interaction Between AJ and TJ in Diseases

The barrier function of the TJ helps in the prevention of passing of inflammatory and cancer cells (Campbell et al. 2017). Upregulation or downregulation of the TJ proteins is considered as one of the reasons for the loss of cell-cell adhesion, laying the path for the cancer or inflammatory cells to proliferate (Bhat et al. 2018). Decreased hepatocyte growth factor (HGF)-mediated trans-epithelial resistance and increased paracellular permeability of human breast cancer cell lines MDA-MB-231 and MCF-7 are examples of TJ dysfunction (Martin and Jiang 2009). While the AJs aid TJs by providing a platform for their assembly, AJ and TJ interactions are associated with a variety of diseases. For example, AJ proteins regulate cell proliferation through Wnt signaling where β-catenin, a transcription factor and a downstream effector in the Wnt signaling pathway, promotes transcription of genes involved in cell proliferation and differentiation. The mutations in the proteins involved in Wnt signaling pathway are associated with dysregulated cellular functions as seen in intestine and colorectal cancer (Garcia et al. 2018). Similarly, E-cadherin regulates the Hippo pathway, which suppresses the activation of the YAP, resulting in decreased cell proliferation. Evidence from previous studies has shown the role of endothelial cadherin in upregulating the TJ protein claudin-5 (Dejana et al. 2009). E-cadherin extracellular domain (E-cad-ECD) functionalized supported lipid bilayer substrates have been used for reconstituting E-cadherin-mediated adhesion. Such hybrid live-cell supported lipid bilayer substrates help to study the recruitment of the α-catenin to particular cell membrane regions and the role of actomyosin contraction in junction formation (Biswas et al. 2015c). The effects of the inflammatory milieu on the function of the TJ proteins and their distribution are much known, while there are fewer studies explaining and examining the function and the effects on AJs (Mehta et al. 2015). E-cadherin or β-catenin expression in intestinal epithelial cells treated with IFN-γ is one such example (Mehta et al. 2015; Aust et al. 2001). These suggests more comprehensive understanding and investigation of AJ barrier function and dysfunction.

2.4.1 Tumor Microenvironment (TME)

The tumor microenvironment is the cellular environment surrounding the tumor cell. It is a complex structure constituting different types of cells embedded in a modified extracellular matrix (ECM), signaling molecules, blood vessels, non-invaded cells, endothelial cells, tumor vasculature, and lymphatics (Balkwill et al. 2012) interacting bidirectionally with the endothelium and mesothelium membranes of the cells and the ECM macromolecules and the tumor cells determining the tumor progression and metastatic dissemination (Salvador et al. 2016). Tumor cells attack the basal membrane and extracellular matrix (ECM) through aberrant proliferation and affect the normal adjacent cells by disrupting their status quo. It is not just the cancer cells that cause the disease; instead, the normal cells recruited to the TME that contribute to cancer or cancer-related symptoms (Vasioukhin 2012). This effect is

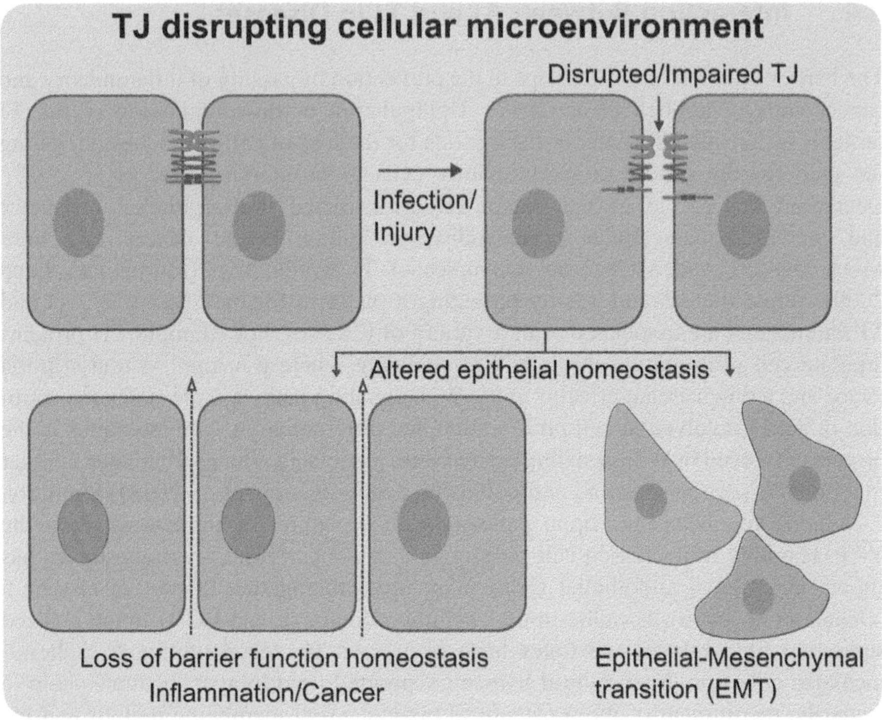

Fig. 2.4 Dysregulation of TJ results altering cellular microenvironment. Impaired TJ resulting from disease or injury leads to the EMT. Disruption in TJs interrupts the barrier function and exposes cells to the external environment leading to altered epithelial homeostasis and cellular adhesion. This loss of barrier function and cellular adhesion accelerates the epithelial-mesenchymal transition in normal cells and metastasis in cancer cells

because of the dysregulation of transmembrane or signaling proteins and the transcription factors that control cell proliferation (Quail and Joyce 2013). Any alteration in the expression of these proteins changes the polarity, increases cell permeability, and the loss of contact inhibition. This phenomenon can be observed in highly metastatic cancer cells, which poorly express TJ proteins (Martin and Jiang 2009).

Epithelial cells are sensitive to the external environment caused by injury or toxin, microbial and viral attack (Christiansen and Rajasekaran 2006). Thus, the cells develop a repair mechanism to prevent cell migration and proliferation in the damaged area. Mesenchymal characteristics are acquired by the epithelial cells during EMT, which causes loss of cell polarity and E-cadherin function (Fig. 2.4) (Salvador et al. 2016; Thiery 2002). Therefore, TJs play their role in the organization of the epithelial layer.

2.4.2 Dysregulation of TJ and AJ in Tumor Microenvironment and Inflammation

Dysregulation of TJs and alteration in the tumor microenvironment are simultaneously involved in cancer cell proliferation and tumorigenesis. Cancer manifestation is generally associated with mutations and inflammation. Yet, TJ dysfunction and functional loss of E-cadherin can also be considered a reason for the onset of disease (Albini and Sporn 2007). Epithelial cells in normal conditions are strongly adhered to each other, while the adhesion between the tumor cells varies immensely. Especially in metastasis, tumor cells lose the cell-cell adhesion and spread throughout the body (Campbell et al. 2017). Decrease in adhesion function and the intercellular adhesion weakening lead to the change in the microenvironment and inflammation. In addition to the loss of E-cadherin expression, the catenins also play a role in the dysregulation of the microenvironment (Garcia et al. 2018). Most breast carcinomas display loss of α-catenin, and β-catenin that can cause tumor formation (Berx and Van Roy 2001). The stability of the AJ mostly depends upon the efficient formation of the cadherin-catenin complexes, and the phosphorylation of these complexes may influence their binding affinities and cause the dysregulation of AJ (Aaltomaa et al. 1999). Additionally, receptor tyrosine kinase activity is known to lead to a decline in the cadherin-catenin complex assembly. The serine/threonine phosphorylation also regulates the stability of the AJs (Vasioukhin 2012). Phosphorylation of β-catenin disrupts the AJ formation. In addition to these adhesion proteins, the Rho-family GTPases also regulate cell-cell adhesion, cell migration, and invasion. The continuous activation of Rho and Rac mostly destabilizes the cadherin-catenin association and stimulates cell migration, tumor formation, and invasion (Vasioukhin 2012).

There is a significant increase in oxidative stress during inflammation because of cytokine activation. Cytokines enhance the inflammatory process causing tumor cell proliferation due to antitumor immunity suppression. They mediate changes in the paracellular permeability that may lead to a number of diseases including inflammatory bowel disease (IBD), cystic fibrosis, and inflammation of airway in asthma (Capaldo and Nusrat 2009). Cytokines intervene with the TJ barrier function through the perturbation of the TJs resulting in increased paracellular permeability (Capaldo and Nusrat 2009). Endothelial and epithelial permeability affected by IFN-γ is yet to be understood at the molecular level but as such the endothelial and epithelial cells react to the IFN-γ treatment by differentially localizing adhesion-associated scaffolding protein while in the inflammatory cells, the transmembrane proteins claudin, occludin, and junction adhesion molecules are internalized away from the cell-cell adhesion regions (Utech et al. 2006). Thus, TME with disrupted TJs would influence tumor metastasis because of uncontrolled cellular permeability (Quante et al. 2013). A deeper understanding of TJ dysregulation can provide a better insight into the mechanisms involved in controlling the TJs and open new therapeutic ways for cancer.

2.4.3 TJ Proteins and Their Role Associated with Cancer

Cancer is a set of heterogeneous conditions and arises through a set of complex process. The phenomenon of cell-cell communication during tumorigenesis is yet not understood and is being explored. One speculated event of cancer is the pathogenic or anomalous stimulus leading to chronic inflammation, which alters the microenvironment of the cells (Salvador et al. 2016; Brücher and Jamall 2014). Metastasis is the primary cause of death in cancer patients. Tight junctions are associated with cell signalling processes and play a role in cell differentiation, growth and in maintaining cell polarity (Itoh and Bissell 2003). In addition to this regulation of cell proliferation by tight junctions helps in suppressing and preventing cells during tumorigenesis Salvador et al. 2016. TJ proteins such as claudins, occludin, tricellulin, ZO, and JAMs which comprise the molecular makeup of tight junction have shown to have a potential role in metastasis and cancer Takano et al. 2014. For example, it has been reported that the inhibition of Raf-1 signaling by occludin induces tumor growth (Wang et al. 2005). Tumorigenic and metastatic properties promoted within the cancer cells is because of the epigenetic silencing of occludin protein. Similarly, the downregulation or decreased occludin protein levels are associated with the metastatic increase in many types of cancers such as breast cancer, liver carcinoma, and ovarian cancer (Salvador et al. 2016).

The transmembrane protein claudin plays a vital role in TJ selectivity (Tsukita et al. 2008). Claudins also mediate the epithelial to mesenchymal transition (EMT). Their role in tumor progression and gene-specific retention in the TJ is known from previous studies (Suh et al. 2013). Various types of cancers can be differentiated based on the differential expression of distinct types of claudins. Claudins 1, 3, 4, and 7 are dysregulated frequently in a multitude of cancers (Morin 2005) while declined expression of claudins 1, 2, and 7 is associated with breast cancer (Salvador et al. 2016).

Both epithelial and endothelial cells as well as platelets express junctional adhesion molecules. They play a role in the metastasis of various cancers where the upregulation of JAM-A induces EMT, and the process of EMT is reversed when the JAM expression is silenced (Bazzoni 2003). Maintenance and proper assembly of the TJ is carried out by the interaction of the transmembrane proteins with ZO-1 (Martin 2014). ZO-1 plays a role in cell proliferation, epithelial tissue morphogenesis, and tumor invasion-associated EMT. A decrease in the ZO proteins increases the risk of invasion of cancers, e.g., breast and colorectal cancers (McNeil et al. 2006).

2.4.3.1 TJs and Regulation of Its Proteins in Different Types of Cancers

Expression of TJ proteins is shown to be dysregulated in different tumorigenic tissues and cancers (Table 2.1). Progression of glioma, which is the most common central nervous system tumor, leads to structural changes in endothelial cells resulting in enhanced permeability and brain edema (Liebner et al. 2000). The TJ proteins claudin-1 and claudin-5 are downregulated, causing tumor progression in human glioma cancer (Morin 2005). Human cutaneous squamous carcinoma (SCC) represents the claudin-4 and ZO-1 protein localization. The other TJ proteins,

Table 2.1 Expression of the TJ proteins in different types of cancer

#	TJ protein	Expression	Cancer type	References
1.	Claudin-1 & claudin-5	Downregulated	Human glioma cancer	Salvador et al. (2016), Liebner et al. (2000)
2.	Claudin-4	Localization	Human cutaneous squamous carcinoma (SCC)	Yang et al. (2015)
3.	Claudin-3	Overexpression	Breast cancer	Morin (2005)
4.	Claudin-1, 4, & 7	Decreased expression	Colorectal cancer	Singh et al. (2011)
5.	Claudin-15	Downregulation	Mega-intestine	Tamura et al. (2008)
Occludin: The loss of occludin at the TJ is the common feature among many types of cancers				
6.	ZO proteins 1 & 2	Low expression	Breast cancer	Fanning et al. (2012)
7.	ZO-1	Localization	Human cutaneous squamous carcinoma (SCC)	Morita et al. (2004)
8.	ZO-1	Downregulation	Gastric cancer	Salvador et al. (2016), Morita et al. (2004)

claudin-1, is downregulated while occludin is completely lost at the TJs. The loss of occludin at the TJ is the common feature among many cancers. Apoptosis is caused by decreased epithelial adhesion due to the knockdown of occludin (Rachow et al. 2013). Tumor progression is mediated by the downregulation of claudin-1 and upregulation of claudin-2 (Rachow et al. 2013). A breast cancer study also showed that the overexpression of claudin-3 disrupted the TJ integrity leading to tumor progression. On the other hand, the lower expression of the ZO proteins 1 and 2 was related to the poor prognosis in breast cancer. The role of occludin expression in maintaining the TJ integrity in the breast tissue is also important. Varied expression of the protein may trigger breast cancer (Salvador et al. 2016).

2.4.3.2 Mutations Affecting TJs

Tight junctions, AJs, and desmosomes are intricately related to each other in a number of tissues and any mutations in the genes encoding their component proteins can potentially result in diseases conditions. Previous studies have provided much evidence about complex cell-cell communication and helped to determine the genes that affect and encode the proteins resulting from mutations (Wilcox et al. 2001). Mutations are significant reasons for the onset of many diseases, including cancer and other genetic disorders (Lai-Cheong et al. 2007). Tight junctions and intercellular communications are also influenced and affected by mutations leading to different types of developmental changes, altered phenotypes, and at times novel phenotypical and genotypical changes (Lai-Cheong et al. 2007). To date, mutations in four TJ proteins (claudin 1, 12, 16, and ZO-2), one AJ protein (P-cadherin), and eight desmosome components have been listed (Konrad et al. 2006).

Mutations in the human claudin genes were the first to be identified. It was followed by the detection of one nonsense and few missense mutations in the claudin-16 associated with the autosomal renal disorder (Lai-Cheong et al. 2007). Mutations in the claudin 14 genes resulted in the autosomal recessive deafness disorder, while the dermatologically inherited TJ mutations are caused by claudin1 disorders (Wilcox et al. 2001; Lai-Cheong et al. 2007).

Mutation or promoter hypermethylation or deletion of E-cadherin results in prostrate, ovarian, lung, and hepatocellular carcinomas (Caldeira et al. 2006). In diffuse gastric cancer (DGC) and lobular breast cancer (LBC), E-cadherin inactivation is considered as an early initiating event. In contrast, its downregulation of E-cadherin in the prostate, lung, and ovarian cancer is regarded as a late event that causes increased invasiveness of cancer cells (Chen et al. 2014). The identification of mutations encoding proteins contributing to the intercellular junctions has facilitated understanding and studying the wide spectrum of inherited diseases.

2.5 Conclusion

In summary, cell-cell contacts are crucial for tissue homeostasis, cell differentiation, and organ integrity. Though the classical view defines AJ and TJ as distinct cellular components, recent evidence delineates their intricate relations and interdependency. Several studies show that AJs and TJs are interconnected physically by ZO-1 via the formation of catenin and afadin protein complexes and, in turn, play a central role in regulating cytoskeleton dynamics. In addition to physical connections, they couple downstream signaling pathways to maintain cell polarity. Polarity of proteins or complexes maintains the interplay between AJs and TJs by regulating the maturation of embryonic AJs and controlling TJ localization. Although substantial developments have been made in understanding the role of TJ and its associated proteins in various diseases and pathology in recent years, the intercellular physical interaction between E-cadherin and TJ transmembrane proteins is poorly explained. Dysregulation in claudins, such as upregulation of ZEB-1 by claudin-1, diminishes the expression of E-cadherin in cancers (Singh et al. 2011). Thus, changes in the expression of claudin makes it an attractive diagnostic marker and widen the scope of understanding the underlying mechanism causing these diseases. The mutations associated with the TJs causing physiological and genetic disorders are yet to be explored. Understanding these processes and studying the mechanism underlying the interaction between AJ and TJ have significant importance including the opening up of novel therapeutic and diagnostic approaches.

References

Aaltomaa S et al (1999) Alpha-catenin expression has prognostic value in local and locally advanced prostate cancer. Br J Cancer 80(3–4):477–482

Albini A, Sporn MB (2007) The tumour microenvironment as a target for chemoprevention. Nat Rev Cancer 7(2):139–147

Altamash T et al (2021) Intracellular ionic strength sensing using NanoLuc. Int J Mol Sci 22(2):677
Aust DE et al (2001) Altered distribution of beta-catenin, and its binding proteins E-cadherin and APC, in ulcerative colitis-related colorectal cancers. Mod Pathol 14(1):29–39
Balkwill FR, Capasso M, Hagemann T (2012) The tumor microenvironment at a glance. J Cell Sci 125(pt 23):5591–5596
Bauer H et al (2010) The dual role of zonula occludens (ZO) proteins. J Biomed Biotechnol 2010: 402593
Bazzoni G (2003) The JAM family of junctional adhesion molecules. Curr Opin Cell Biol 15(5): 525–530
Berx G, Van Roy F (2001) The E-cadherin/catenin complex: an important gatekeeper in breast cancer tumorigenesis and malignant progression. Breast Cancer Res 3(5):289–293
Bhat AA et al (2018) Tight junction proteins and signaling pathways in cancer and inflammation: a functional crosstalk. Front Physiol 9:1942
Biswas KH (2020) Molecular mobility-mediated regulation of E-cadherin adhesion. Trends Biochem Sci 45(2):163–173
Biswas KH, Groves JT (2016) A microbead supported membrane-based fluorescence imaging assay reveals intermembrane receptor-ligand complex dimension with Nanometer precision. Langmuir 32(26):6775–6780
Biswas KH, Zaidel-Bar R (2017) Early events in the assembly of E-cadherin adhesions. Exp Cell Res 358(1):14–19
Biswas KH et al (2015a) E-cadherin junction formation involves an active kinetic nucleation process. Proc Natl Acad Sci U S A 112(35):10932–10937
Biswas KH et al (2015b) Cyclic nucleotide binding and structural changes in the isolated GAF domain of Anabaena adenylyl cyclase, CyaB2. PeerJ 3:e882
Biswas KH et al (2015c) E-cadherin junction formation involves an active kinetic nucleation process. Proc Natl Acad Sci 112(35):10932–10937
Biswas KH et al (2016) Sustained alpha-catenin activation at E-cadherin junctions in the absence of mechanical force. Biophys J 111(5):1044–1052
Biswas KH et al (2018) Multicomponent supported membrane microarray for monitoring spatially resolved cellular signaling reactions. Adv Biosyst 2(4):1800015
Boyer B, Vallés AM, Edme N (2000) Induction and regulation of epithelial-mesenchymal transitions. Biochem Pharmacol 60(8):1091–1099
Brücher BL, Jamall IS (2014) Cell-cell communication in the tumor microenvironment, carcinogenesis, and anticancer treatment. Cell Physiol Biochem 34(2):213–243
Caldeira JR et al (2006) CDH1 promoter hypermethylation and E-cadherin protein expression in infiltrating breast cancer. BMC Cancer 6:48
Campbell HK, Maiers JL, DeMali KA (2017) Interplay between tight junctions & adherens junctions. Exp Cell Res 358(1):39–44
Capaldo CT, Macara IG (2007) Depletion of E-cadherin disrupts establishment but not maintenance of cell junctions in Madin-Darby canine kidney epithelial cells. Mol Biol Cell 18(1):189–200
Capaldo CT, Nusrat A (2009) Cytokine regulation of tight junctions. Biochim Biophys Acta 1788(4):864–871
Chen A et al (2014) E-cadherin loss alters cytoskeletal organization and adhesion in non-malignant breast cells but is insufficient to induce an epithelial-mesenchymal transition. BMC Cancer 14: 552
Chen Z et al (2018) Spatially modulated ephrinA1:EphA2 signaling increases local contractility and global focal adhesion dynamics to promote cell motility. Proc Natl Acad Sci U S A 115(25): E5696–E5705
Chen Z et al (2021) Probing the effect of clustering on EphA2 receptor signaling efficiency by subcellular control of ligand-receptor mobility. elife 10:e67379
Christiansen JJ, Rajasekaran AK (2006) Reassessing epithelial to mesenchymal transition as a prerequisite for carcinoma invasion and metastasis. Cancer Res 66(17):8319–8326

Contreras RG et al (2002) E-cadherin and tight junctions between epithelial cells of different animal species. Pflugers Arch 444(4):467–475
Cummins PM (2012) Occludin: one protein, many forms. Mol Cell Biol 32(2):242–250
Davis MA, Ireton RC, Reynolds AB (2003) A core function for p120-catenin in cadherin turnover. J Cell Biol 163(3):525–534
Dejana E et al (2009) Organization and signaling of endothelial cell-to-cell junctions in various regions of the blood and lymphatic vascular trees. Cell Tissue Res 335(1):17–25
Ebnet K (2017) Junctional adhesion molecules (JAMs): cell adhesion receptors with pleiotropic functions in cell physiology and development. Physiol Rev 97(4):1529–1554
Fanning AS, Van Itallie CM, Anderson JM (2012) Zonula occludens-1 and -2 regulate apical cell structure and the zonula adherens cytoskeleton in polarized epithelia. Mol Biol Cell 23(4): 577–590
Furman C et al (2007) Ena/VASP is required for endothelial barrier function in vivo. J Cell Biol 179(4):761–775
Garcia MA, Nelson WJ, Chavez N (2018) Cell-cell junctions organize structural and signaling networks. Cold Spring Harb Perspect Biol 10(4):a029181
González-Mariscal L, Betanzos A, Ávila-Flores A (2000) MAGUK proteins: structure and role in the tight junction. Semin Cell Dev Biol 11(4):315–324
Harris TJC, Tepass U (2010) Adherens junctions: from molecules to morphogenesis. Nat Rev Mol Cell Biol 11(7):502–514
Hartsock A, Nelson WJ (2008) Adherens and tight junctions: structure, function and connections to the actin cytoskeleton. Biochim Biophys Acta 1778(3):660–669
Hotz B et al (2007) Epithelial to mesenchymal transition: expression of the regulators snail, slug, and twist in pancreatic cancer. Clin Cancer Res 13(16):4769–4776
Ikenouchi J et al (2007) Requirement of ZO-1 for the formation of belt-like adherens junctions during epithelial cell polarization. J Cell Biol 176(6):779–786
Itoh M, Bissell MJ (2003) The organization of tight junctions in epithelia: implications for mammary gland biology and breast tumorigenesis. J Mammary Gland Biol Neoplasia 8 (4):449–462
Ivanov DB, Philippova MP, Tkachuk VA (2001) Structure and functions of classical cadherins. Biochemistry (Mosc) 66(10):1174–1186
Kabir HB, Nam Joon C, Jay TG (2018) Fabrication of multicomponent, spatially segregated DNA and protein-functionalized supported membrane microarray. Langmuir 34:9781–9788
Kallakury BV, Sheehan CE, Ross JS (2001) Co-downregulation of cell adhesion proteins alpha- and beta-catenins, p120CTN, E-cadherin, and CD44 in prostatic adenocarcinomas. Hum Pathol 32(8):849–855
Knights AJ et al (2012) Holding tight: cell junctions and cancer spread. Trends Cancer Res 8:61–69
Konrad M et al (2006) Mutations in the tight-junction gene claudin 19 (CLDN19) are associated with renal magnesium wasting, renal failure, and severe ocular involvement. Am J Hum Genet 79(5):949–957
Krause G et al (2008) Structure and function of claudins. Biochim Biophys Acta 1778(3):631–645
Kyuno D et al (2014) Targeting tight junctions during epithelial to mesenchymal transition in human pancreatic cancer. World J Gastroenterol 20(31):10813–10824
Kyuno D et al (2021) Role of tight junctions in the epithelial-to-mesenchymal transition of cancer cells. Biochim Biophys Acta 1863(3):183503
Lai-Cheong JE, Arita K, McGrath JA (2007) Genetic diseases of junctions. J Investig Dermatol 127(12):2713–2725
Lamouille S, Xu J, Derynck R (2014) Molecular mechanisms of epithelial-mesenchymal transition. Nat Rev Mol Cell Biol 15(3):178–196
Liebner S et al (2000) Claudin-1 and claudin-5 expression and tight junction morphology are altered in blood vessels of human glioblastoma multiforme. Acta Neuropathol 100(3):323–331
Lioni M et al (2007) Dysregulation of claudin-7 leads to loss of E-cadherin expression and the increased invasion of esophageal squamous cell carcinoma cells. Am J Pathol 170(2):709–721

Maiers JL et al (2013) ZO-1 recruitment to α-catenin--a novel mechanism for coupling the assembly of tight junctions to adherens junctions. J Cell Sci 126(Pt 17):3904–3915
Martin TA (2014) The role of tight junctions in cancer metastasis. Semin Cell Dev Biol 36:224–231
Martin TA, Jiang WG (2009) Loss of tight junction barrier function and its role in cancer metastasis. Biochim Biophys Acta 1788(4):872–891
McNeil E, Capaldo CT, Macara IG (2006) Zonula occludens-1 function in the assembly of tight junctions in Madin-Darby canine kidney epithelial cells. Mol Biol Cell 17(4):1922–1932
Mège RM, Ishiyama N (2017) Integration of cadherin adhesion and cytoskeleton at adherens junctions. Cold Spring Harb Perspect Biol 9(5):a028738
Mehta S et al (2015) Defects in the adherens junction complex (E-cadherin/β-catenin) in inflammatory bowel disease. Cell Tissue Res 360(3):749–760
Meng W, Takeichi M (2009) Adherens junction: molecular architecture and regulation. Cold Spring Harb Perspect Biol 1(6):a002899–a002899
Miyaguchi K (2000) Ultrastructure of the zonula adherens revealed by rapid-freeze deep-etching. J Struct Biol 132(3):169–178
Morin PJ (2005) Claudin proteins in human cancer: promising new targets for diagnosis and therapy. Cancer Res 65(21):9603–9606
Morita K, Tsukita S, Miyachi Y (2004) Tight junction-associated proteins (occludin, ZO-1, claudin-1, claudin-4) in squamous cell carcinoma and Bowen's disease. Br J Dermatol 151(2):328–334
Ozawa M, Kobayashi W (2014) Cadherin cytoplasmic domains inhibit the cell surface localization of endogenous E-cadherin, blocking desmosome and tight junction formation and inducing cell dissociation. PLoS One 9(8):e105313–e105313
Quail DF, Joyce JA (2013) Microenvironmental regulation of tumor progression and metastasis. Nat Med 19(11):1423–1437
Quante M et al (2013) The gastrointestinal tumor microenvironment. Gastroenterology 145(1): 63–78
Rachow S et al (2013) Occludin is involved in adhesion, apoptosis, differentiation and Ca2+–homeostasis of human keratinocytes: implications for tumorigenesis. PLoS One 8(2):e55116
Rasool S, Geethakumari AM, Biswas KH (2021) Role of actin cytoskeleton in E-cadherin-based cell–cell adhesion assembly and maintenance. J Indian Inst Sci 101(1):51–62
Rübsam M et al (2017) E-cadherin integrates mechanotransduction and EGFR signaling to control junctional tissue polarization and tight junction positioning. Nat Commun 8(1):1250
Salvador E, Burek M, Förster CY (2016) Tight junctions and the tumor microenvironment. Curr Pathobiol Rep 4:135–145
Scott JA et al (2006) Ena/VASP proteins can regulate distinct modes of actin organization at cadherin-adhesive contacts. Mol Biol Cell 17(3):1085–1095
Shigetomi K et al (2018) Adherens junctions influence tight junction formation via changes in membrane lipid composition. J Cell Biol 217(7):2373–2381
Shin K, Fogg VC, Margolis B (2006) Tight junctions and cell polarity. Annu Rev Cell Dev Biol 22(1):207–235
Singh AB et al (2011) Claudin-1 up-regulates the repressor ZEB-1 to inhibit E-cadherin expression in colon cancer cells. Gastroenterology 141(6):2140–2153
Soini Y (2011) Claudins in lung diseases. Respir Res 12(1):70
Suh Y et al (2013) Claudin-1 induces epithelial-mesenchymal transition through activation of the c-Abl-ERK signaling pathway in human liver cells. Oncogene 32(41):4873–4882
Swaney KF, Li R (2016) Function and regulation of the Arp2/3 complex during cell migration in diverse environments. Curr Opin Cell Biol 42:63–72
Takano K et al (2014) Role of tight junctions in signal transduction: an update. EXCLI J 13:1145–1162
Tamura A et al (2008) Megaintestine in claudin-15-deficient mice. Gastroenterology 134(2): 523–534
Thiery JP (2002) Epithelial-mesenchymal transitions in tumour progression. Nat Rev Cancer 2(6): 442–454

Tsai JH, Yang J (2013) Epithelial-mesenchymal plasticity in carcinoma metastasis. Genes Dev 27(20):2192–2206

Tsukita S et al (2008) Tight junction-based epithelial microenvironment and cell proliferation. Oncogene 27(55):6930–6938

Twiss F et al (2012) Vinculin-dependent cadherin mechanosensing regulates efficient epithelial barrier formation. Biol Open 1(11):1128–1140

Utech M, Brüwer M, Nusrat A (2006) Tight junctions and cell-cell interactions. Methods Mol Biol 341:185–195

Vasioukhin V (2012) Adherens junctions and cancer. In: Harris T (ed) Adherens junctions: from molecular mechanisms to tissue development and disease. Springer, Dordrecht, pp 379–414

Wang Z et al (2005) The second loop of occludin is required for suppression of Raf1-induced tumor growth. Oncogene 24(27):4412–4420

Wilcox ER et al (2001) Mutations in the gene encoding tight junction claudin-14 cause autosomal recessive deafness DFNB29. Cell 104(1):165–172

Yang Y et al (2015) Interleukin-18 enhances breast cancer cell migration via down-regulation of claudin-12 and induction of the p38 MAPK pathway. Biochem Biophys Res Commun 459(3): 379–386

Zemljic-Harpf AE et al (2014) Vinculin directly binds zonula occludens-1 and is essential for stabilizing connexin-43-containing gap junctions in cardiac myocytes. J Cell Sci 127(Pt 5): 1104–1116

Tight Junctions, Epithelial-Mesenchymal Transition, and Cancer Metastasis

3

Roohi Mohi-ud-din, Rafia Jan, Inamu Rashid Khan, Sheema Hashem, Rashid Mir, Imadeldin Elfaki, Tariq Masoodi, Shahab Uddin, Muzafar A. Macha, and Ajaz A. Bhat

Abstract

Epithelial-mesenchymal transition (EMT) induces dynamic changes in the cellular organization leading to functional changes in cell migration, invasion, and metastasis. Neoplastic cells with specific genetic and epigenetic alterations

R. Mohi-ud-din · R. Jan
Department of Pharmaceutical Sciences, University of Kashmir, Srinagar, Jammu and Kashmir, India

I. R. Khan
Department of Zoology, School of Life Sciences, Central University of Kashmir, Ganderbal, Jammu and Kashmir, India

S. Hashem
Department of Human Genetics, Sidra Medicine, Doha, Qatar

R. Mir
Department of Medical Lab Technology, Prince Fahad Bin Sultan Chair for Biomedical Research, Faculty of Applied Medical Sciences, University of Tabuk, Tabuk, Saudi Arabia

I. Elfaki
Department of Biochemistry, Prince Fahad Bin Sultan Chair for Biomedical Research, Faculty of Applied Medical Sciences, University of Tabuk, Tabuk, Saudi Arabia

T. Masoodi
Human Immunology Department, Research Branch, Sidra Medicine, Doha, Qatar

S. Uddin
Translational Research Institute, Academic Health System, Hamad Medical Corporation, Doha, Qatar

M. A. Macha
Watson-Crick Centre for Molecular Medicine, Islamic University of Science and Technology, Awantipora, Jammu and Kashmir, India

A. A. Bhat (✉)
Department of Human Genetics-Precision Medicine in Diabetes, Obesity and Cancer Research Program, Sidra Medicine, Doha, Qatar
e-mail: abhat@sidra.org

A. A. Bhat et al. (eds.), *Tight Junctions in Inflammation and Cancer*,
https://doi.org/10.1007/978-981-99-2415-8_3

undergo EMT. While the epithelial cell comprises various classes of cell junctions, including desmosomes, adherens junctions (AJ), tight junctions (TJ), and gap junctions (GJ), TJs are considered central players in regulating epithelial cell function. The TJ proteins include junctional adhesion molecules (JAM), Pals1 (Proteins Associated with Lin Seven 1), cingulin, MUPP1 (multi-PDZ domain protein 1), tricellulin, and ZO1, ZO-2, ZO-3 (Zona occludens). Claudins have been found to have abnormal expressions in a variety of malignancies. Differential expression of TJ proteins has been observed in various malignancies, including prostate, lung, breast, ovarian, esophageal, colorectal, and gastric cancers. Disruption in the expression of TJ proteins results in the EMT phenotype associated with loss of cell-cell adhesion and dissociation of cells from primary tumor followed by invasion and metastasis to distant organs. In addition, claudins and occludins, the essential TJ proteins, also mediate various crucial functions in the cell. TJ proteins are also hepatocyte entry factors for the hepatitis-C virus (HCV), a leading cause of liver disease and cancer worldwide. This chapter comprehensively discusses the current knowledge on the involvement of TJ proteins in the EMT and metastasis of various cancers.

Keywords

Cancer · Cell proliferation · Claudins · Epithelial-mesenchymal transition · Invasion · Metastasis · Tight junctions

3.1 Introduction

An epithelial-mesenchymal transition (EMT) is a process that allows a polarized epithelial cell to undergo various biochemical changes that would enable it to adopt the mesenchymal cell phenotype with greater migratory capacity, invasiveness, and excessive synthesis of extracellular matrix (ECM) components (Kalluri and Neilson 2003). Degradation of the underlying basement membrane and the production of mesenchymal cells that may move away from the epithelial layer in which they originated signify the end of EMT (Kalluri and Neilson 2003). EMTs are found in three different biological environments, each with its functional ramifications. The class of EMTs known as "type 1" does not promote fibrosis nor induce an invasive phenotype that leads to systemic dissemination across the circulatory system (Kalluri and Neilson 2003). Type 1 EMTs can produce mesenchymal cells (primary mesenchyme) with the ability to undergo a mesenchymal-epithelial transition (MET) and generate secondary epithelia, among other aspects. A second type of EMT is linked to wound healing, tissue regeneration, and organ fibrosis. The program starts in this type 2 EMTs as part of a repair-related process that ordinarily generates fibroblasts and other related cells to rebuild tissues following trauma and inflammatory injury. Type 2 EMTs are related to inflammation and stop when the inflammation subsides, as seen during wound healing and tissue regeneration (Marconi et al. 2021). Type 2 EMTs can continue to respond to persistent inflammation in the presence of organ fibrosis, eventually leading to organ loss. Type 3 EMTs are seen in neoplastic cells that have undergone earlier genetic and epigenetic alterations, particularly in genes that promote clonal expansion and the establishment of confined malignancies. These alterations, which mostly affect oncogenes and tumor suppressor genes, combine with the EMT regulatory circuitry to create results that differ significantly from those seen in the other two kinds of EMT. Carcinoma cells that undergo type 3 EMT have the potential to penetrate and metastasize, resulting in the final, life-threatening signs of cancer growth. Notably, cancer cells may undergo EMTs to varying degrees, with some cells keeping many epithelial characteristics while acquiring some mesenchymal features and others shedding all epithelial markers and becoming entirely mesenchymal. What particular cues cause type 3 EMTs in cancer cells is still unknown. Such signals could, for example, arise in the tumor stroma, which is linked to many primary carcinomas (Kalluri and Weinberg 2009). Many steps of the invasion-metastasis cascade are completed by carcinoma cells with EMT-induced mesenchymal traits, such as local invasion of neoplastic cells at the primary tumor site, intravasation into blood vessels, translocation through the circulation, extravasation into the parenchyma of distant tissues, and survival as micrometastatic deposits (Lambert et al. 2017). An epithelial cell comprises various classes of cell junctions, including desmosomes, adherens junctions (AJ), tight junctions (TJ), and gap junctions (GJ). TJ and AJ are the leading players in regulating epithelial cell function (Kyuno et al. 2021). TJ or Zonula occludens represent the extremes of the epithelial and endothelial cells, distinguishing the upper and lower surface and establishing polarity to these cells (Bhat et al. 1942; Matter and Balda 2003). TJ represents that section of the plasma membrane where the adjoining cells form a sequence of connections that obstruct the extracellular space, establishing a barrier between the cells and creating an intramembrane diffusion wall (Wong and Gumbiner 1997). TJs preserve the cell's polarity and govern the passage of substances across the epithelial cells. TJs maintain cellular polarity by building a barrier that restricts the apical/basolateral passage of lipids in the outer leaflet of the plasma membrane (Matter and Balda 2003).

Moreover, they participate in various cellular functions such as differentiation, migration, and proliferation by interacting with cytoskeletal elements and downstream signaling molecules. TJs maintain the selective passage of multiple ions, water molecules, macromolecules, and immune cells within the plasma membrane. The dysregulation and disruption of TJs have been implicated in various disease conditions including cancer.

3.2 Tight Junction Proteins

Tight junctions are places where tissues interface directly with the external environment or internal compartments that are contiguous with the external environment and are lined by mucosal surfaces, where epithelial cells act as internal organ insulation. These structures serve as a protective covering and a selective barrier between the body and the gut lumen, preventing free movement across the

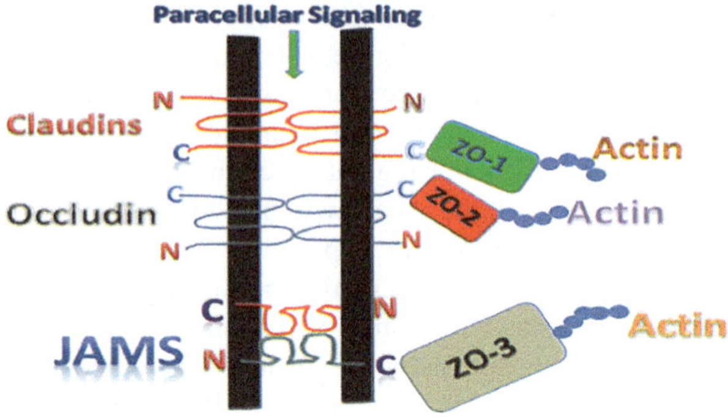

Fig. 3.1 The three types of tight junction proteins (Claudin and Occludin are tetraspanin membrane proteins, JAM is a single membrane-spanning protein). Zonula occludin (ZO) proteins cover each molecule's N & C terminus and are associated with actin filaments

paracellular region (Farquhar and Palade 1963). Intercellular junctions, or TJs, are made up of various proteins. TJ proteins recruit signaling proteins in response to external stimuli, which aids critical regulatory cell activities, including differentiation, proliferation, and migration. TJ proteins are also hepatocyte entry factors for the hepatitis C virus (HCV), a leading cause of liver disease and cancer worldwide. Chronic HCV infection, cholestatic liver disorders, and hepatobiliary cancer have all been linked to changes in TJ protein expression (Roehlen et al. 2020). Claudins and occludins are the essential tight junction proteins that mediate various crucial functions in the cell. The other tight junctional proteins include junctional adhesion molecules (JAM), Pals1 (Proteins Associated with Lin Seven 1), cingulin, MUPP1 (multi-PDZ domain protein 1) tricellulin, and ZO1, ZO-2, ZO-3 (Zona occludens) (Bhat et al. 2019; Gowrikumar et al. 2019). These proteins interact between themselves and with the cytoskeleton to generate a complex architecture (Fig. 3.1) (Gowrikumar et al. 2019).

3.3 Claudin Proteins

Tsukita and his colleagues discovered claudins as critical integral membrane proteins in 1998, before which occludin was the sole TJ protein reported. Claudins are prevalent in both epithelia and endothelia and generate paracellular barriers and pores that govern the permeability of TJ. They are recognized as essential structural components of TJ (Günzel and Yu 2013). Claudins are tetraspan proteins that range in size from 18 to 27 kDa and are made up of four transmembrane domains: cytoplasmic N terminus (7 amino acids), cytoplasmic C terminus (25–55 amino acids), and two extracellular loops (ECLs). They have no sequence homology with occludin. ECL1 and ECL2 are two extracellular loops with 25 and 50 amino acids, respectively (Chiba et al. 2008). Claudins have been found to have abnormal expressions in a variety of malignancies. Claudin-1, -3, -4, and -7 are some claudins that are often dysregulated in malignancies (Morin 2005). For the well-documented role of claudins in cancer stem cell control, a link between medication resistance and distant metastasis seems unavoidable and clear (Wang et al. 2019).

3.4 Metastasis

Metastasis is the movement of cancer cells to tissues or organs other than the tissues where cancer initially emerged and results in the development of new tumors at different sites (secondary and tertiary foci) (Ribatti et al. 2020). Invasion, intravasation, and extravasation are the primary events that make up the metastatic cascade. Malignant tumor cells detach from the initial tumor mass due to a loss of cell-cell adhesion capacity, and alterations in cell-matrix interaction allow the cells to infiltrate the surrounding stroma; this is the invasion process. This includes the secretion of chemicals that destroy the basement membrane and extracellular matrix and the expression/suppression of proteins that regulate motility and migration. Local diffusion for delivery of nutrients to and removal of waste products from the tumor site would be adequate for a tumor up to 2 mm in diameter, but without angiogenesis, the tumor would fail to develop (Brooks 1996). The climax of tumor cell evolution may be distant metastasis. Individual cancer cells or a subgroup of cancer cells manage to break through the physical boundaries of a primary tumor and complete a series of rate-limiting stages to populate secondary sites (Welch and Hurst 2019). Over the last few decades, much progress has been made in the field of diagnosis and cancer therapeutics; however, cancer metastasis remains a significant obstacle in achieving favorable clinical outcomes and accounts for 90% of cancer-related fatalities (Mehlen and Puisieux 2006). Metastasis is a highly flawed process; however, conventional therapeutic approaches such as radiotherapy, chemotherapy, and targeted therapy have proven to be successful in curbing cancer growth but still are responsible for large percentage of cancer-related mortalities due to the resistance to these conventional treatments (Shen and Kang 2020). During metastasis, cancer cells are put to the extreme "survival of the fittest" test since only a tiny proportion of dispersed tumor cells can conquer various obstacles, including immunological and therapeutic attacks. Metastasis is a highly complex process that involves numerous steps. The initial step is invasion, characterized by loss of cellular connections and increased cell motility that conquers the surrounding tissues. In the intravasation phase of metastasis, tumor cells cross the endothelium of blood and lymph vessels and reach systemic circulation. However, only a small percentage of circulating tumor cells can make it through circulation and undergo a process of extravasation, where these cells escape through the capillary and reach distal sites (Yang et al. 2004). The time between organ infiltration and colonization is known as metastatic latency. Even though metastatic latency is highly related to metastatic occurrences, little information is known due to the scarcity of experimental models (Massagué and Obenauf 2016). The competence of infiltrated tumor cells determines how long this period lasts, and it is subject to many physiological limitations. Furthermore, this mechanism varies depending on the type of cancer. The ability of tumor cells to spread after decades of dormancy, as shown in breast and prostate malignancies, indicates that competence can be acquired over time in such circumstances (Lee 1985).

3.5 Tight Junctions and Cancer Metastasis

Tight junctions maintain cell polarity by establishing a barrier restricting the membrane lipids and proteins to diffuse laterally, thus sustaining the distinct composition of apical and basolateral domains. Further, the involvement of junction proteins in activating various signaling cascades suggested their role in the different cellular functions, including cell differentiation and proliferation (Morin 2005). Disruption in the expression of TJ proteins results in loss of cell-cell adhesion that promotes dissociation of cells from primary tumor followed by invasion and eventually leading to metastasis to distant organs. Moreover, loss of cell-cell adhesion in TJ impairs cellular polarity that induces the dedifferentiation of cancer cells' characteristics. Numerous previous studies have demonstrated that the loss of various junctional proteins, such as occludin and claudin, can accelerate cancer progression. However, this perspective has been contraindicated by multiple previous studies suggesting that increased expression of various TJ proteins contributed to cancer growth and progression (Leech et al. 2015). Therefore, optimum junction protein levels are vital to preserving normal body functioning, and any imbalance may lead to severe pathological implications. The expression of various tight junction proteins is altered in different malignancies (Table 3.1).

3.6 Epithelial to Mesenchymal Transition

Epithelia are considered as single or multicellular tissue that serves different purposes. Epithelial cells exhibit apical-basal polarity and adhere and interact with one another across unique intercellular junctions (Huang et al. 2012; Lee and Streuli 2014). The EMT is characterized by loss of junctional integrity and apicobasal polarity of epithelial cells and acquires motile mesenchymal phenotype (Fig. 3.1). The mesenchymal characteristic of the cell involves cancer cells' greater invasive and migratory capacity, resistance to apoptosis and chemotherapy, and a marked increase in the production of extracellular matrix proteins (ECM) (Kalluri and Weinberg 2009; Skrypek et al. 2017; Mittal 2018; Moreno-Bueno et al. 2008). The EMT is responsible for the progression and metastasis of tumor cells and is also engaged in the generation of tumor cells that exhibit stem cell features that contribute to developing resistance to chemotherapy treatment (Roche 2018). The epithelial state of the cells in which EMT is initiated is defined by stable cell-cell junctions, apicobasal polarity, and interaction with the basement membrane (Yang et al. 2020). The accomplishment of EMT is marked by the disintegration of the basement membrane and the emergence of mesenchymal cells that may move away from its epithelial layer (Kalluri and Weinberg 2009). Further, during EMT, the epithelial cells demonstrate upregulation of mesenchymal markers such as fibronectin, vimentin, N-cadherin, and fibroblast-specific protein-1, and downregulation of epithelial markers such as occluding and ZO-1 (Mittal 2018). However, EMT plays an essential role in many physiological functions, such as wound healing and embryogenesis but is abnormally stimulated in many pathological circumstances,

Table 3.1 Tight junction proteins in different malignancies

Junctional protein type	Cancer type	Mechanism of action	Ref.
Occludin	Lung cancer	Suppressing the expression of occludin inhibits the metastatic potential by inhibiting proliferation and promoting apoptosis of various lung cancer cell lines	Wang et al. (2018)
		Epigenetic silencing of the occludin gene increases the tumorigenic, invasive, and metastatic potential of cancer cells through the regulation of various genes that are associated with apoptosis	Osanai et al. (2006)
	Breast cancer (MDA-MB-241) cell line	Loss in the expression of occludin resulted in an increase in the metastasis to bone	Martin et al. (2016)
Claudin-1		Claudin-1 increases the invasive potential of oral squamous cell carcinoma cells by prompting the cleavage of the laminin-5 gamma2 chain mediated by metalloproteinase (MMP)-2 and MMP-1	Oku et al. (2006)
Claudin-2	Breast cancer	Claudin-2 encourages the liver metastasis of breast cancer cells by enabling seeding and early cancer cell survival	Tabariès et al. (2021)
Claudin-3 and Claudin-4	Ovarian cancer	Overexpression of Claudin-3 and Claudin-4 in epithelial cells of the ovary encourages tumorigenesis and metastasis by enhancing cancer cell survival and invasion	Agarwal et al. (2005)
Claudin 4	Gastric cancer	Overexpression of Claudin-4 increases the metastasis of AGS cells by increasing the expression of metalloproteinase-2 and 9	Hwang et al. (2014)
	Colonic cancer	Increased expression of claudin-4 increases the invasiveness by increasing expression of MMP-2 and 9	Takehara et al. (2009)
JAM-A	Renal cancer (Renal cell carcinoma (RCC) cell line)	Reduced expression of JAM-A was found to be associated with the establishment of renal cancer and enhances the migratory property of these cells, which might be due to the activation of metalloproteinases	Gutwein et al. (2009)
	Colorectal cancer	Downregulation of JAM-A by microRNA 21 enhances the progression, migratory, and metastatic potential of colorectal cancer by activating the pro-survival and pro-migratory signaling cascade	Gutwein et al. (2009)
	Breast cancer	Overexpression in the JAM-A protein in the MDA-MB-231 breast cancer cell line restricted the migration and invasion of these cell lines. Further, the cell lines that migrated the least were found to have expressed high levels of JAM-A protein.	Gutwein et al. (2009)

(continued)

Table 3.1 (continued)

Junctional protein type	Cancer type	Mechanism of action	Ref.
		The anti-migratory property attributed to JAM-A is due to formation and strengthening of tight junctions	
	Breast cancer	Suppression of JAM-A by TGF-α increases the invasive property of the MDA-MB-231 cell line by inhibiting the transcription of the JAM-A gene	Wang and Lui (2012)
	Nasopharyngeal cancer	JAM-A increases metastasis and invasiveness in human nasopharyngeal cancer by inducing EMT through activation of P13K/Akt signaling cascade	Tian et al. (2015)
	Gastric cancer	Lower expression of JAM-A molecules was associated with poor clinical outcomes and enhancement in the invasive and metastatic properties in these patients	Huang et al. (2014)
ZO-1	Liver cancer	Overexpression of ZO-1 prevented the cancer cell metastasis by preventing cancer cell viability, proliferation, and migration and induction of G0/G1 cell cycle arrest	Zhang et al. (2019)
	Colorectal cancer	Decreased expression of ZO-1 resulted in liver metastasis	Kaihara et al. (2003)

such as fibrosis and cancer cell growth (Lee and Streuli 2014; Their 2002). Various transcriptional factors mediate the triggering of EMT by suppressing the epithelial gene expression while increasing mesenchymal gene expression through multiple mechanisms that are currently not well known (Skrypek et al. 2017). The hallmarks of the mesenchymal phenotype are increased expression of N-cadherin, Fibronectin, Tenascin C, Vimentin, collagen VI α, and laminin β1 (Książkiewicz et al. 2012; Lai et al. 2020). In contrast, E-cadherin, cytokeratins, and occludins are fundamental and most critical characteristic markers of epithelial state (Lai et al. 2020).

Various molecular mechanisms are involved that lead to the initiation of EMT. These include recruitment of various transcription factors, increase in expression of different cell surface proteins, expression of cytoskeletal proteins and their reorganization, synthesis of multiple enzymes responsible for degradation of ECM, and altered expression of certain micro-RNAs (Fig. 3.2) (Kalluri and Weinberg 2009). The activation of various transcription factors such as SNAIL, zinc-finger E-box-binding (ZEB), TWIST1/2, and basic helix-loop-helix and reprogramming in gene expression is regulated by multiple signaling, among which transforming growth factor-β (TGFβ) family has a crucial role (Lamouille et al. 2014; Jung et al. 2015). The other signaling cascade involved in EMT includes Notch, fibroblast, NF kappa B, and Wnt pathways (Moreno-Bueno et al. 2008; Ramos et al. 2017). Further, the fibroblast growth factor (FGF), hepatocyte growth factor (HGF), and

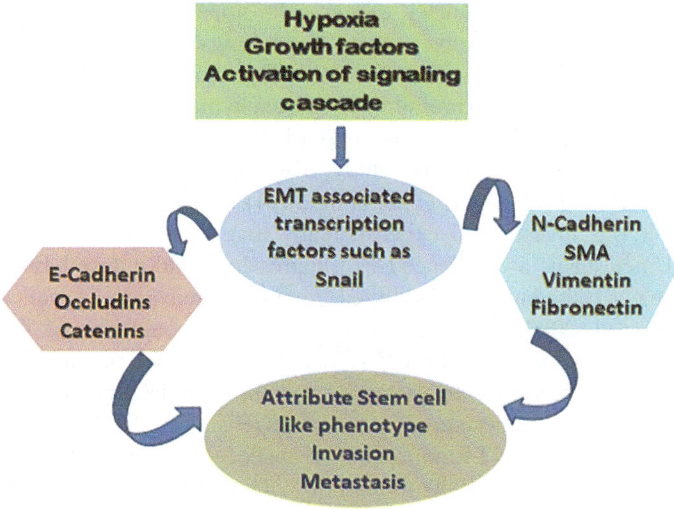

Fig. 3.2 Epithelial to mesenchymal transition. Hypoxia induces hypoxia-inducible factor (HIF) stabilization and downstream signaling, which has pleiotropic roles in cancers. The hallmarks of the mesenchymal phenotype are increased expression of N-cadherin, Fibronectin, and Vimentin. Activating transcription factors such as SNAIL also have a crucial role in EMT. Moreover, E-Cadherin and Occludin expression is also associated with metastasis and invasion, which play a crucial role in EMT

epidermal growth factor (EGF) are also known to regulate EMT (Lai et al. 2020; Nisticò et al. 2012). These signaling cascades activate transcription factors that suppress the genes that encode for epithelial characteristics such as E-cadherin and activate transcriptional processes that define invasive mesenchymal behavior (Moreno-Bueno et al. 2008). A wide variety of microenvironmental factors promote EMT, such as pro-inflammatory cytokinin produced locally by activated stromal cells, ECM proteins such as fibronectin, collagen-1 and Hyaluronan, and hypoxia (Jung et al. 2015). Further noncoding RNAs (ncRNAs), chromatin remodeling, and epigenetic alterations are also responsible for the transition from EMT invasive phenotype (De Craene and Berx 2013). In addition to its prominent involvement in encouraging tumor cell invasion, EMT imparts resistance to these cells to apoptosis and anoikis, therefore permitting cell survival of the tumor cells in the bloodstream following intravasation (Jung et al. 2015).

3.7 EMT and Cancer Metastasis

The role of EMT in causing cancer metastasis was illustrated by examining the expression of various EMT transcription factors such as Twist1, Snail1, and Prrx1 (Mittal 2018). A previous study showed that Twist1 expression was related to the induction of metastatic characteristics. The reduction in Twist expression in highly invasive and metastatic mammary cancer cells impedes the metastasis of these cancerous cells from the mammary gland to the lungs (Yang et al. 2004). Further, aberrant expression of Twist results in loss of E-cadherin-mediated cell-cell adhesion, enhancement in the expression of mesenchymal markers, and increase in cellular motility, which suggests the role of Twist in metastasis by promoting EMT (Yang et al. 2004). In another study, it was observed that genetic ablation of Snail1 inhibited lung metastasis (Tran et al. 2014).

Further, it was investigated experimentally that snail-induced EMT facilitated cancer cell metastasis, as evidenced by suppressing snail-induced EMT not only prevented metastasis but also inhibited immunosuppression in cancer patients (Kudo-Saito et al. 2009). Previous studies revealed that ablation of Zeb1, a transcription factor and activator of EMT in a mouse model of pancreatic cancer, resulted in a 60% decrease in metastatic potential, indicating that Zeb1-driven EMT may contribute to metastasis in cancer (Krebs et al. 2017). Inhibition in the expression of Twist in the 4T1 mouse breast cancer cell line prevented the metastatic potential of these carcinoma cells to the lungs. This confers that Twist, a transcriptional regulator of EMT, has a key role in causing metastasis. Downregulation of Twist1 in oral squamous cell carcinoma (OSCC) decreased their metastatic potential in both in vivo and in vitro studies (da Silva et al. 2014). Further suppression in the expression level of TWIST1 by thymoquinone prevented metastasis in various cancer cell lines such as MDA-MB-435, HeLa, and BT549 (Khan et al. 2015). However, there are some conflicting results regarding the association between metastasis and EMT.

3.8 Conclusion

TJs serve a key function in maintaining cell polarity by establishing a barrier restricting the membrane lipids and proteins from diffusing laterally. Disruption in the expression of TJ proteins results in loss of cell-cell adhesion that promotes dissociation of cells from primary tumor followed by invasion and eventually leading to metastasis to distant organs. Chronic HCV infection, cholestatic liver disorders, and hepatobiliary cancer have all been linked to changes in TJ protein expression. Among TJ proteins, Claudins have been found to have abnormal expressions in various malignancies (Ding et al. 2013). The EMT is characterized by loss of junctional integrity and acquiring motile mesenchymal phenotype. The mesenchymal characteristic of the cell involves greater migratory capacity, increased invasive ability of cancer cells, and development of resistance to apoptotic cell death and chemotherapy. Although EMT plays an important role in many physiological functions, such as during wound healing and embryogenesis but is abnormally stimulated in many pathological circumstances, such as fibrosis and cancer cell growth. The hallmarks of the mesenchymal phenotype are increased expression of N-cadherin, Fibronectin, Tenascin C, Vimentin, collagen VI α, and laminin $\beta1$. Metastasis is the movement of cancer cells to tissues or organs other than the tissues where cancer initially emerged and resulted in the development of new tumors at different sites. Invasion, intravasation, and extravasation are the primary events that make up the metastatic cascade. Over the last few decades, much progress has been made in the field of diagnosis and cancer therapeutics; however, cancer metastasis remains a significant obstacle to achieving favorable clinical outcomes. Although conventional therapeutic approaches such as radiotherapy, chemotherapy, and targeted therapy have proven successful in curbing cancer growth, many cancer-related mortalities are still due to resistance to these conventional treatments.

Acknowledgments The authors would like to acknowledge the Sidra Medicine Precision Program for funding and constant support to Ajaz A. Bhat. This study was also supported by Ramalingaswami Re-entry Fellowship (Grant number: D.O. NO.BT/HRD/35/02/2006) from the Department of Biotechnology, Govt. of India, New Delhi, and & Core Research Grant (CRG/2021/003805) from Science and Engineering Research Board (SERB), Govt. of India, New Delhi, to Muzafar A. Macha.

Declarations **Ethical Approval and Consent to Participate**: Not Applicable.

Consent for Publication Not Applicable.

Availability of Supporting Data Not Applicable.

Competing Interests The authors declare that they have no competing interests.

References

Agarwal R, D'Souza T, Morin PJ (2005) Claudin-3 and claudin-4 expression in ovarian epithelial cells enhances invasion and is associated with increased matrix metalloproteinase-2 activity. Cancer Res 65(16):7378–7385

Bhat AA, Uppada S, Achkar IW, Hashem S, Yadav SK, Shanmugakonar M, Al-Naemi HA, Haris M, Uddin S (1942) Tight junction proteins and signaling pathways in cancer and inflammation: a functional crosstalk. Front Physiol 2018:9

Bhat AA, Uppada S, Achkar IW, Hashem S, Yadav SK, Shanmugakonar M, Al-Naemi HA, Haris M, Uddin S (2019) Tight junction proteins and signaling pathways in cancer and inflammation: a functional crosstalk. Front Physiol 9:1942

Brooks PC (1996) Cell adhesion molecules in angiogenesis. Cancer Metastasis Rev 15(2):187–194

Chiba H, Osanai M, Murata M, Kojima T, Sawada N (2008) Transmembrane proteins of tight junctions. Biochim Biophys Acta 1778(3):588–600

De Craene B, Berx G (2013) Regulatory networks defining EMT during cancer initiation and progression. Nat Rev Cancer 13(2):97–110

Ding L, Lu Z, Lu Q, Chen YH (2013) The claudin family of proteins in human malignancy: a clinical perspective. Cancer Manage Res 5:367–375

Farquhar MG, Palade GE (1963) Junctional complexes in various epithelia. J Cell Biol 17(2): 375–412

Gowrikumar S, Singh AB, Dhawan P (2019) Role of claudin proteins in regulating cancer stem cells and chemoresistance-potential implication in disease prognosis and therapy. Int J Mol Sci 21(1): 53

Günzel D, Yu ASL (2013) Claudins and the modulation of tight junction permeability. Physiol Rev 93(2):525–569

Gutwein P, Schramme A, Voss B, Abdel-Bakky MS, Doberstein K, Ludwig A, Altevogt P, Hansmann ML, Moch H, Kristiansen G et al (2009) Downregulation of junctional adhesion molecule-A is involved in the progression of clear cell renal cell carcinoma. Biochem Biophys Res Commun 380(2):387–391

Huang RY, Guilford P, Thiery JP (2012) Early events in cell adhesion and polarity during epithelial-mesenchymal transition. J Cell Sci 125(pt 19):4417–4422

Huang JY, Xu YY, Sun Z, Wang ZN, Zhu Z, Song YX, Luo Y, Zhang X, Xu HM (2014) Low junctional adhesion molecule A expression correlates with poor prognosis in gastric cancer. J Surg Res 192(2):494–502

Hwang T-L, Changchien T-T, Wang C-C, Wu C-M (2014) Claudin-4 expression in gastric cancer cells enhances the invasion and is associated with the increased level of matrix metalloproteinase-2 and -9 expression. Oncol Lett 8(3):1367–1371

Jung HY, Fattet L, Yang J (2015) Molecular pathways: linking tumor microenvironment to epithelial-mesenchymal transition in metastasis. Clin Cancer Res 21(5):962–968

Kaihara T, Kusaka T, Nishi M, Kawamata H, Imura J, Kitajima K, Itoh-Minami R, Aoyama N, Kasuga M, Oda Y et al (2003) Dedifferentiation and decreased expression of adhesion molecules, E-cadherin and ZO-1, in colorectal cancer are closely related to liver metastasis. J Exp Clin Cancer Res 22(1):117–123

Kalluri R, Neilson EG (2003) Epithelial-mesenchymal transition and its implications for fibrosis. J Clin Invest 112(12):1776–1784

Kalluri R, Weinberg RA (2009) The basics of epithelial-mesenchymal transition. J Clin Invest 119(6):1420–1428

Khan MA, Tania M, Wei C, Mei Z, Fu S, Cheng J, Xu J, Fu J (2015) Thymoquinone inhibits cancer metastasis by downregulating TWIST1 expression to reduce epithelial to mesenchymal transition. Oncotarget 6(23):19580–19591

Krebs AM, Mitschke J, Lasierra Losada M, Schmalhofer O, Boerries M, Busch H, Boettcher M, Mougiakakos D, Reichardt W, Bronsert P et al (2017) The EMT-activator Zeb1 is a key factor for cell plasticity and promotes metastasis in pancreatic cancer. Nat Cell Biol 19(5):518–529

Książkiewicz M, Markiewicz A, Zaczek AJ (2012) Epithelial-mesenchymal transition: a hallmark in metastasis formation linking circulating tumor cells and cancer stem cells. Pathobiology 79(4):195–208

Kudo-Saito C, Shirako H, Takeuchi T, Kawakami Y (2009) Cancer metastasis is accelerated through immunosuppression during Snail-induced EMT of cancer cells. Cancer Cell 15(3): 195–206

Kyuno D, Takasawa A, Kikuchi S, Takemasa I, Osanai M, Kojima T (2021) Role of tight junctions in the epithelial-to-mesenchymal transition of cancer cells. Biochim Biophys Acta 1863(3): 183503

Lai X, Li Q, Wu F, Lin J, Chen J, Zheng H, Guo L (2020) Epithelial-mesenchymal transition and metabolic switching in cancer: lessons from somatic cell reprogramming. Front Cell Dev Biol 8: 760

Lambert AW, Pattabiraman DR, Weinberg RA (2017) Emerging biological principles of metastasis. Cell 168(4):670–691

Lamouille S, Xu J, Derynck R (2014) Molecular mechanisms of epithelial-mesenchymal transition. Nat Rev Mol Cell Biol 15(3):178–196

Lee YT (1985) Patterns of metastasis and natural courses of breast carcinoma. Cancer Metastasis Rev 4(2):153–172

Lee J, Streuli C (2014) Integrins and epithelial cell polarity. J Cell Sci 127:3217

Leech AO, Cruz RGB, Hill ADK, Hopkins AM (2015) Paradigms lost-an emerging role for over-expression of tight junction adhesion proteins in cancer pathogenesis. Ann Transl Med 3(13): 184–184

Marconi GD, Fonticoli L, Rajan TS, Pierdomenico SD, Trubiani O, Pizzicannella J, Diomede F (2021) Epithelial-mesenchymal transition (EMT): the type-2 EMT in wound healing, tissue regeneration and organ fibrosis. Cells 10(7):1587

Martin TA, Jordan N, Davies EL, Jiang WG (2016) Metastasis to bone in human cancer is associated with loss of occludin expression. Anticancer Res 36(3):1287

Massagué J, Obenauf AC (2016) Metastatic colonization by circulating tumour cells. Nature 529(7586):298–306

Matter K, Balda MS (2003) Functional analysis of tight junctions. Methods 30(3):228–234

Mehlen P, Puisieux A (2006) Metastasis: a question of life or death. Nat Rev Cancer 6(6):449–458

Mittal V (2018) Epithelial mesenchymal transition in tumor metastasis. Annu Rev Pathol 13:395–412

Moreno-Bueno G, Portillo F, Cano A (2008) Transcriptional regulation of cell polarity in EMT and cancer. Oncogene 27(55):6958–6969

Morin PJ (2005) Claudin proteins in human cancer: promising new targets for diagnosis and therapy. Cancer Res 65(21):9603–9606

Nisticò P, Bissell MJ, Radisky DC (2012) Epithelial-mesenchymal transition: general principles and pathological relevance with special emphasis on the role of matrix metalloproteinases. Cold Spring Harb Perspect Biol 4(2):a011908

Oku N, Sasabe E, Ueta E, Yamamoto T, Osaki T (2006) Tight junction protein claudin-1 enhances the invasive activity of oral squamous cell carcinoma cells by promoting cleavage of laminin-5 gamma2 chain via matrix metalloproteinase (MMP)-2 and membrane-type MMP-1. Cancer Res 66(10):5251–5257

Osanai M, Murata M, Nishikiori N, Chiba H, Kojima T, Sawada N (2006) Epigenetic silencing of occludin promotes tumorigenic and metastatic properties of cancer cells via modulations of unique sets of apoptosis-associated genes. Cancer Res 66(18):9125–9133

Ramos FS, Wons L, Cavalli IJ, Ribeiro E (2017) Epithelial-mesenchymal transition in cancer: an overview. Integr Cancer Sci Ther 4:1–5

Ribatti D, Tamma R, Annese T (2020) Epithelial-mesenchymal transition in cancer: a historical overview. Transl Oncol 13(6):100773

Roche J (2018) The epithelial-to-mesenchymal transition in cancer. Cancers 10(2):52

Roehlen N, Roca Suarez AA, El Saghire H, Saviano A, Schuster C, Lupberger J, Baumert TF (2020) Tight junction proteins and the biology of hepatobiliary disease. Int J Mol Sci 21(3):825

Shen M, Kang Y (2020) Stresses in the metastatic cascade: molecular mechanisms and therapeutic opportunities. Genes Dev 34(23–24):1577–1598

da Silva SD, Alaoui-Jamali MA, Soares FA, Carraro DM, Brentani HP, Hier M, Rogatto SR, Kowalski LP (2014) TWIST1 is a molecular marker for a poor prognosis in oral cancer and represents a potential therapeutic target. Cancer 120(3):352–362

Skrypek N, Goossens S, De Smedt E, Vandamme N, Berx G (2017) Epithelial-to-mesenchymal transition: epigenetic reprogramming driving cellular plasticity. Trends Genet 33(12):943–959

Tabariès S, Annis MG, Lazaris A, Petrillo SK, Huxham J, Abdellatif A, Palmieri V, Chabot J, Johnson RM, Van Laere S et al (2021) Claudin-2 promotes colorectal cancer liver metastasis and is a biomarker of the replacement type growth pattern. Commun Biol 4(1):657

Takehara M, Nishimura T, Mima S, Hoshino T, Mizushima T (2009) Effect of claudin expression on paracellular permeability, migration and invasion of colonic cancer cells. Biol Pharm Bull 32(5):825–831

Their J (2002) Epithelial-mesenchymal transitions in tumor progression. Nat Rev Cancer 2(6): 442–454

Tian Y, Tian Y, Zhang W, Wei F, Yang J, Luo X, Zhou T, Hou B, Qian S, Deng X et al (2015) Junctional adhesion molecule-A, an epithelial-mesenchymal transition inducer, correlates with metastasis and poor prognosis in human nasopharyngeal cancer. Carcinogenesis 36(1):41–48

Tran HD, Luitel K, Kim M, Zhang K, Longmore GD, Tran DD (2014) Transient SNAIL1 expression is necessary for metastatic competence in breast cancer. Cancer Res 74(21): 6330–6340

Wang Y, Lui WY (2012) Transforming growth factor-β1 attenuates junctional adhesion molecule-A and contributes to breast cancer cell invasion. Eur J Cancer 48(18):3475–3487

Wang M, Liu Y, Qian X, Wei N, Tang Y, Yang J (2018) Downregulation of occludin affects the proliferation, apoptosis and metastatic properties of human lung carcinoma. Oncol Rep 40(1): 454–462

Wang K, Li T, Xu C, Ding Y, Li W, Ding L (2019) Claudin-7 downregulation induces metastasis and invasion in colorectal cancer via the promotion of epithelial-mesenchymal transition. Biochem Biophys Res Commun 508(3):797–804

Wong V, Gumbiner BM (1997) A synthetic peptide corresponding to the extracellular domain of occludin perturbs the tight junction permeability barrier. J Cell Biol 136(2):399–409

Welch DR, Hurst DR (2019) Defining the hallmarks of metastasis. Cancer Res 79(12):3011–3027

Yang J, Mani SA, Donaher JL, Ramaswamy S, Itzykson RA, Come C, Savagner P, Gitelman I, Richardson A, Weinberg RA (2004) Twist, a master regulator of morphogenesis, plays an essential role in tumor metastasis. Cell 117(7):927–939

Zhang X, Wang L, Zhang H, Tu F, Qiang Y, Nie C (2019) Decreased expression of ZO-1 is associated with tumor metastases in liver cancer. Oncol Lett 17(2):1859–1864

Yang J, Antin P, Berx G, Blanpain C, Brabletz T, Bronner M, Campbell K, Cano A, Casanova J, Christofori G et al (2020) Guidelines and definitions for research on epithelial–mesenchymal transition. Nat Rev Mol Cell Biol 21(6):341–352

Imaging Techniques to Study Tight Junctions

4

Tayyiba Akbar Ali, Sabah Akhtar, Sabah Nisar, Tariq Masoodi, Ravinder Reddy, Ajaz A. Bhat, and Mohammad Haris

Abstract

Tight junctions (TJs) are important regulators of paracellular permeability that act as protective barriers against pathogens and maintain cell polarity and homeostasis by regulating the flow of ions, solutes, and macromolecules. TJs are essential to the barrier integrity of epithelial and endothelial cells, and the loss of their function is linked to various pathological conditions. To better understand the molecular mechanisms underlying the junctional architecture, it is vital to understand the nanoscale organization of the molecular components of TJs, such as occludins, junctional adhesion molecules (JAMs), and, most importantly

Tayyiba Akbar Ali and Sabah Akhtar contributed equally as first authors.

T. A. Ali
Division of Genomics and Translational Medicine, College of Health and Life Sciences, Hamad Bin Khalifa University, Doha, Qatar

S. Akhtar
Translational Research Institute, Academic Health System, Hamad Medical Corporation, Doha, Qatar

S. Nisar
Department of Diagnostic Imaging, St. Jude Children's Research Hospital, Memphis, TN, USA

T. Masoodi
Laboratory of Cancer Immunology and Genetics, Sidra Medicine, Doha, Qatar

R. Reddy · M. Haris (✉)
Center for Advanced Metabolic Imaging in Precision Medicine, Department of Radiology, Perelman School of Medicine, University of Pennsylvania, Philadelphia, PA, USA
e-mail: Mohammad.Haris@pennmedicine.upenn.edu

A. A. Bhat
Department of Human Genetics-Precision Medicine in Diabetes, Obesity and Cancer Research Program, Sidra Medicine, Doha, Qatar

A. A. Bhat et al. (eds.), *Tight Junctions in Inflammation and Cancer*,
https://doi.org/10.1007/978-981-99-2415-8_4

claudins. Molecular imaging techniques such as ultrasound, positron emission tomography (PET), single-photon emission computed tomography (SPECT), and magnetic resonance imaging (MRI) enable the visualization of microstructures at a cellular level. In this chapter, we discuss the significance of molecular imaging modalities and their application in visualizing TJ components in various diseases. We posit that imaging TJs and other junctional adhesion complexes can provide insight into the nanoarchitecture of TJ components in model organisms that can help study the specificity of therapeutic agents involved in modulating the paracellular permeability. Improved knowledge of the function and regulation of TJs using imaging techniques can help in recognizing the underlying mechanisms that lead to compromised barrier integrity and lay the groundwork for developing strategies that target TJs for improved drug delivery with minimal toxicity and facilitate the development of imaging-based biomarkers to enhance the mechanistic understanding of TJ dysfunction in various pathologies.

Keywords

Tight junctions · Molecular imaging · Positron emission tomography · Single-photon emission computed tomography · Magnetic resonance imaging · Ultrasound

4.1 Introduction

Proper cell adhesion is essential to allow accurate signals to be transmitted across the cell and for the mechanical connection of cytoskeletons. The cell-cell adhesion complexes consist of desmosomes, adherens junctions, and tight junctions (Vanslembrouck et al. 2022). Desmosomes and adherens are primarily involved in the firm adhesion of cells, while tight junctions (TJs) provide the fence to regulate the selective paracellular diffusion of molecules based on the size and charge across the cell membrane that is crucial for the tissue's homoeostasis (Zihni et al. 2016; Farquhar and Palade 1963). In the middle of the twentieth century, TJs were discovered using thin-section electron microscopy (Tsukita et al. 2019). They are regions where the membranes of two nearby cells meet and form a barrier (Zihni et al. 2016). TJs are the most enigmatic among all components of cell adhesion. TJs comprise occludin, claudins (a 27-membered protein family), junctional adhesion molecules (JAMs), and Zonula occludens (ZO 1-3) family linker proteins (Verin and Bogatcheva 2006). Occludin, a transmembrane protein, belongs to the MARVEL (MAL and related proteins for vesicle trafficking and membrane connection) protein family, while JAMs belong to the immunoglobulin protein superfamily. Both occludin and JAMs are preferentially situated at the sites where two cells meet, known as bicellular tight junctions (Raleigh et al. 2010). The first two members of claudins were cloned in 1998 that forms a paracellular barrier (Furuse et al. 1998). Claudins have expanded to include 27 members with distinct cell types and tissue

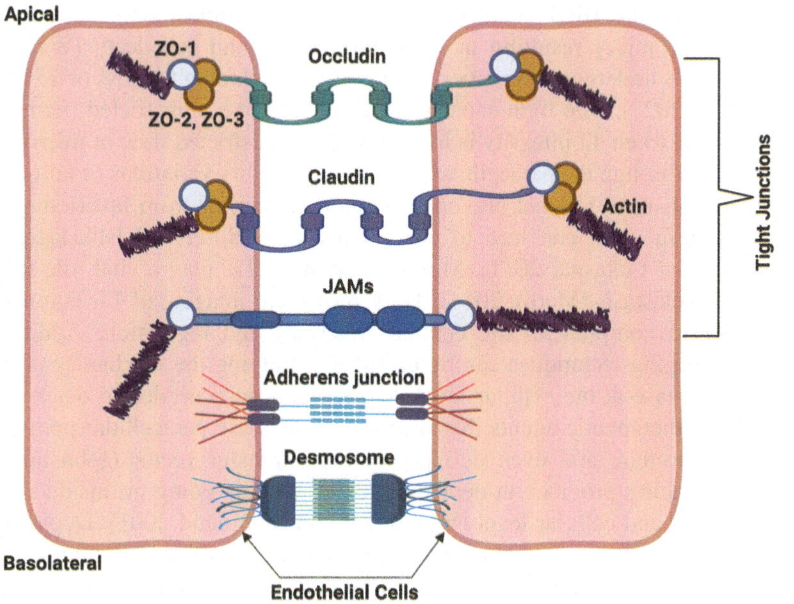

Fig. 4.1 The structure of epithelial junctional complexes. Tight junctions (TJs) are made of transmembrane proteins, claudins, occludins, and junctional adhesion molecules (JAMs) that seal the paracellular gap between epithelial cells. The lower junctional complex, adherens junction are involved in the attachment of adjacent epithelial cells. The junctional complex at the end of the basolateral membrane, desmosomes attach epithelial cells to each other. The figure was created using BioRender

expression patterns (Tsukita et al. 2019). ZO 1-3 are cytoplasmic proteins and interestingly, ZO-1 was the first protein to be linked to the tight junctions that is involved in linking F-actin filaments to the cytoplasmic surface of TJs structures (Gumbiner et al. 1991; Haskins et al. 1998; Jesaitis and Goodenough 1994) (Fig. 4.1).

The length of TJs between two nearby cells is typically 200–500 nm long and 11–15 nm broad, with 10 nm of intermembrane space (Farquhar and Palade 1963). The ultrastructural visualization reveals that TJs are dynamic in nature and extremely ordered meshwork of fibrils and alterations in its composition cause ultrastructural level reorganization (Piontek et al. 2011; Bartosova et al. 2021). Using freeze-fracture electron microscopy, TJs were divided into two pathways depending on the water and solute permeability, i.e., tight and the leaky pathway. The former pathway allows the ions and water movement in size and charge selective manner; however the latter involves movement in a nonselective way (Shen et al. 2011).

TJs generate rows of large, overlapping occlusions between brain microvascular endothelial cells and greatly limit polar molecules and macromolecules paracellular diffusion across blood-brain barrier (BBB) (Lochhead et al. 2020). TJs between endothelial cells keep blood-borne chemicals away from getting into the brain.

However, when the BBB is disrupted, e.g., during intracerebral hemorrhage or traumatic brain injury resulting in vasogenic edema and cell death (Wang et al. 2016), the BBB undergoes a number of alterations, including the loss of TJ proteins (Amoo et al. 2022). The high-capacity and nonselective unrestricted permeability pathway occur when TJ integrity is lost, allowing unrestricted flow of microbes and large proteins through the paracellular space. Permeability variations or failure in the repair mechanism of TJs that prevent the epithelial barrier from functioning accurately is harmful as it can lead to various diseases (Slifer and Blikslager 2020; Bednarczyk and Lukasiuk 2011). More importantly, TJs play a vital role in tumor growth and metastasis (Martin 2014). Thus, molecular imaging of TJs is fundamental in order to comprehend the cellular functions and regulation. Additionally, molecular imaging techniques can be useful for studying the mechanics of human health and disease at the cellular level, and they are also needed to determine the specificity of therapeutic agents that increase or decrease paracellular permeability for peptide, protein, and even delivery of gene transfer vector (Johnson 2005). Molecular imaging provides in-depth visuals of what is going on inside the body on a molecular and cellular level (Weissleder and Mahmood 2001). Depending on the research question, several imaging techniques can be utilized to examine specific proteins, tissue ultrastructure, or the behavior of particular complexes in pathological situations. This chapter delivers the knowledge about some of the important imaging techniques that can be utilized for tight junctions' analysis to provide the best diagnosis or treatment for cancer.

4.2 Positron Emission Tomography

Positron Emission Tomography (PET) is a gold standard in cancer clinics for determination of molecular expression of normal and diseased tissues. PET has frequently been combined with computed tomography (CT) to obtain multimodal anatomic images of a diseased tissue in vivo (Khan et al. 2020). ^{18}F-fluorodeoxyglucose (^{18}F-FDG) has widely been employed as an important PET tracer for glucose metabolism at different stages of various cancer types, and therefore, FDG-PET has emerged as a significant tool for the diagnosis and treatment of cancer (Vander Heiden et al. 2009). As the malignant tumor exhibits high glucose metabolism, the increased FDG uptake by PET helps differentiate between benign and malignant tumors and predict prognosis (Yasuda et al. 2000). Moreover, it has also been employed to determine therapy response after chemotherapy and radiotherapy and also possess the ability to detect residual or recurrent tumor.

Higashi and coworkers reported that low expression of E-cadherin was associated with increased ^{18}F-FDG uptake in human lung adenocarcinoma patients whereas high expression of E-cadherin was associated with decreased ^{18}F-FDG uptake (Higashi et al. 2017). Similarly, in cervical cancer high rad-score of ^{18}F-FDG PET-CT was associated with low E-cadherin expression and low rad-score of ^{18}F-FDG PET-CT indicated high expression of E-cadherin (Li et al. 2021). This indicates the potential association of E-cadherin and ^{18}F-FDG uptake with the tumor stage. Although the uptake of ^{18}F-FDG has not been associated with the tumor size in lung adenocarcinoma, the high ^{18}F-FDG uptake was correlated with short disease-free survival (Higashi et al. 2017). Besides the correlation of ^{18}F-FDG PET-CT rad-score and E-cadherin in cervical cancer, a slight positive correlation of E-cadherin and rad-score with pelvic lymph node metastasis (PLNM) has also been reported (Li et al. 2021). Collectively, these studies demonstrate the utilization of PET imaging to identify tumor metastasis through the expression of E-cadherin.

In addition to E-cadherin, PET imaging has also been employed to study claudin-4 expression in ovarian cancer. Two ^{64}Cu-labeled PET probes for C-terminus of clostridium perfringens enterotoxin protein (cCPE, binds to claudin-4 with high affinity) have been reported for studying ovarian cancer. The two chelators NODAGA-NCS and NODAGA-Mal were synthesized for cCPE. ^{64}Cu-labeled PET probes, NODAGA-NCS and NODAGA-Mal, displayed high uptake in OVCAR3 (claudin-4 positive) cells whereas the uptake in claudin-4 negative ovarian cancer cells SKOV3 was less. Moreover, PET/CT imaging of small animal showed high accumulation of NODAGA-NCS conjugates than NODAGA-Mal conjugates which indicated better in vivo metabolism of 64Cu-cCPE-NODAGA-NCS than 64Cu-cCPE- NODAGA-Mal (Liu et al. 2016). Additionally, in colorectal cancer patients, ^{18}F-FDG PET-CT demonstrated that tumors exhibiting high expression of cadherin-related genes have high transport rate of ^{18}F-FDG (Cheng et al. 2015). Another recent study utilized ^{124}I-5C9 immuno-PET imaging for the detection of claudin-18.2 lesions and near-infrared fluorescent imaging probe (FD1080-5C9) to facilitate the surgical removal of lesions using cancer xenograft models (Zhao et al. 2022) (Fig. 4.2).

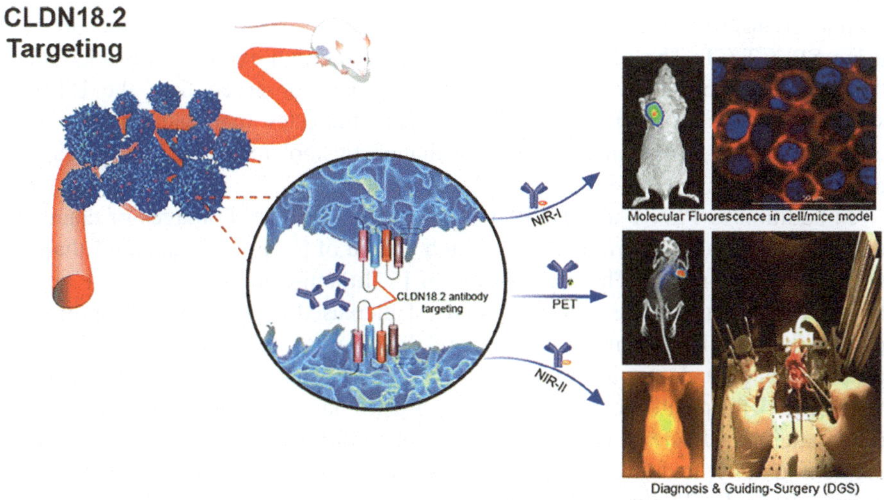

Fig. 4.2 Schematic representation of claudin-18.2 targeting using PET probe ^{124}I-5C9 and near infrared fluorescent imaging probe Cy5.5-5C9 in cancer xenograft model. Reproduced with permission from Molecular Pharmaceutics and Zhao et al. (2022)

4.3 Single-Photon Emission Computed Tomography

Single-photon emission computed tomography (SPECT) has been an integral part of nuclear medicine for many decades as it provides three-dimensional images for radiotracer distribution (Bybel et al. 2008; Madsen 2007). SPECT can detect tumor metastasis; nevertheless, it displays lower sensitivity than PET and is not easily quantifiable (Torigian et al. 2007; Schillaci 2005). Sometimes, SPECT cannot precisely detect the site of disease than uptake radiopharmaceutical; therefore, the combination of SPECT images with anatomical images of CT scans enhances the ability to localize disease and the site of physiological uptake (Schillaci 2004). The combination of functional and anatomical images using SPECT/CT has been increasingly accepted as it helps in identifying underlying pathophysiology and also the understanding of diagnostic information (Schillaci 2005).

In cancer, SPECT/CT has been mainly used to study the expression of claudin-4. Literature shows that high claudin-4 expression in cancer is correlated with high uptake of SPECT radiopharmaceuticals. Claudin-4 expression is associated with distant metastasis and high tumor grade in various cancers such as gastric cancer and breast cancer (Ma et al. 2015; Kim et al. 2022; Hwang et al. 2010, 2014; Jiwa et al. 2014). Moreover, it is also highly expressed in tumor tissues of ovarian cancer, gastric cancer, prostate cancer, and breast cancer than the normal tissues (Tabariès and Siegel 2017). However, studies have also demonstrated that reduced expression of claudin-4 in colorectal carcinoma cells is associated with poor prognosis, progression, metastasis, and histological grade (Ueda et al. 2007).

SPECT/CT has been employed to detect claudin-4 targeted tumor in immunocompromised, xenograft mice. ^{111}In-labeled cCPE peptides were injected in BALB/c nu/nu mice bearing PSN-1 (human pancreatic adenocarcinoma) or HT1080 (human connective tissue epithelial fibrosarcoma) tumor xenografts and then imaged using SPECT/CT. The PSN-1 xenografts expressed high expression of claudin-4; therefore, a high uptake of cCPE radiotracers was observed than the HT1080 xenografts (Baguña Torres et al. 2020). This indicates that claudin-4 targeted cCPE imaging agents can be used as a diagnostic tool for SPECT in pancreatic cancer (Fig. 4.3).

In addition to PSN-1 and HT1080 xenografts, SPECT/CT images of claudin-4 expressing panc-1 xenografts revealed high uptake of [^{111}In]anti-claudin-4 than the uptake of [^{111}In]mIgG. The high uptake of [^{111}In]anti-claudin-4 in panc-1 tumors than HT1080 tumors was also indicative of increased expression of claudin-4 in panc-1 than HT1080 xenografts (Fig. 4.4). Therefore, the uptake of [^{111}In]anti-claudin-4 in panc-1 tumors was mediated by the expression of claudin-4 in these xenografts. This positive association of [^{111}In]anti-claudin-4 uptake and claudin-4 in pancreatic ductal adenocarcinoma was also confirmed by immunofluorescence, autoradiography, and H&E staining (Torres et al. 2018). Additionally, SPECT/CT imaging using [^{125}I]anti-claudin 4 was also employed for Colo357 tumor xenografts of pancreatic cancer in SCID mice. Although the Colo357 cells grow on the orthotopic locations, the findings indicated a high radioactivity in pancreatic tumor

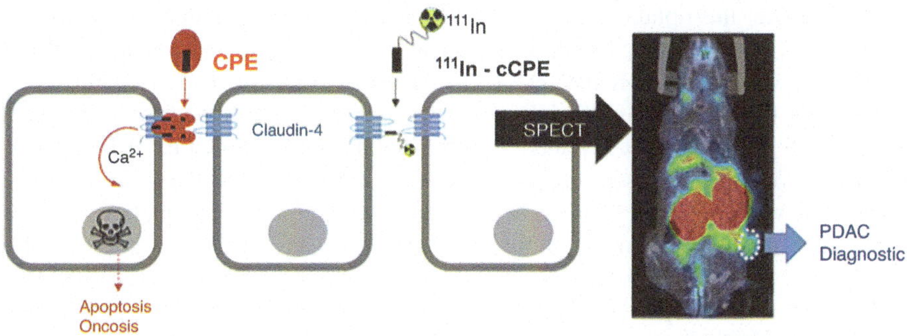

Fig. 4.3 Optimized cCPE-based SPECT imaging agents ([^{111}In]In-cCPE^{L254F+}K257D) as claudin-4-targeting vectors for in vivo imaging of claudin-4 overexpression in pancreatic ductal adenocarcinoma (PDAC). Reproduced with permission from the Journal of Nuclear Medicine and Baguña Torres et al. (2020)

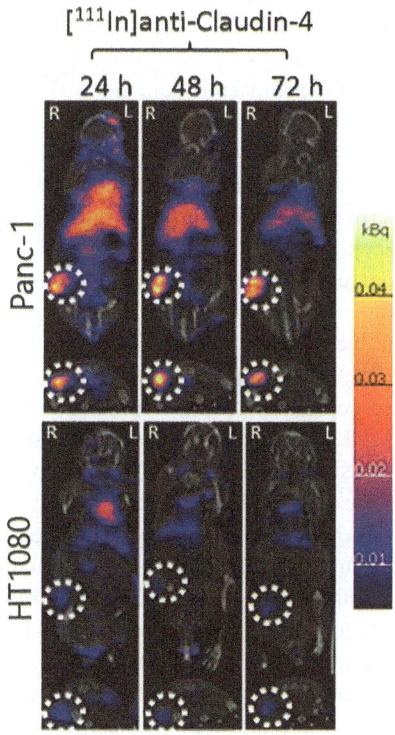

Fig. 4.4 Coronal and transaxial SPECT/CT images of mice bearing Panc-1 (top row) and HT1080 (bottom row) xenograft tumors administered with [^{111}In]anti-claudin-4 intravenously. Tumor uptake of [^{111}In]anti-claudin-4 in Panc-1 xenografts was found to be significantly higher compared to HT1080 tumors. Reproduced with permission from Molecular Imaging and Biology and Torres et al. (2018)

than the spleen and liver of the mice, which reveals that pancreatic cancer exhibits high claudin-4 expression (Foss et al. 2007).

SPECT/CT has also been employed to study the expression of claudin-4 in breast cancer as the imaging has demonstrated that in MDA-MB-468 xenografts of athymic

BALB/c mice, the uptake of the ^{111}In-cCPE.GST was higher than the uptake of ^{111}In-GST. On the other hand, the uptake of ^{111}In-cCPE.GST in the HT1080 xenografts was lower (Mosley et al. 2014). Moreover, the increased uptake of ^{111}In-cCPE.GST in MDA-MB-468 was associated with high claudin-4 expression. SPECT/CT imaging also revealed that the uptake of ^{111}In-cCPE.GST in tumor tissues of BALB/*neuT was higher than the control of* ^{111}In-GST. This is also associated with the high claudin-4 expression in mammary fat pads of BALB/*neuT* mice than the normal mammary fat pads (Mosley et al. 2014, 2015). The expression of claudin-4 is higher at the early stages of tumorigenesis; therefore, ^{111}In-cCPE SPECT permits better visualization of these tumors. Additionally, claudin-4 SPECT imaging can also be used to detect the recurrent cancer.

4.4 Magnetic Resonance Imaging

MRI scans are being used to diagnose a wide range of diseases including cancers. In MRI, the most commonly used paramagnetic contrast agent (CA) is Gadolinium (Gd) which is coupled to a chelating chemical, such as diethylene-triaminepentaacetic acid Gd-DTPA (Knight et al. 2009). The CA helps in measuring the BBB permeability, whose molecules can leak from the intravascular to the interstitial region depending on the degree of BBB breakdown (Raja et al. 2018). When the BBB is breached, such as during multiple sclerosis, brain tumors, and strokes large proteins leak into the brain tissue and cerebrospinal fluid, which can be seen with CA MRI (Raja et al. 2018). However, in some conditions such as dementia, the pathological changes are far less striking, resulting in minor leaks in the neurovascular unit, the series of interfaces between blood and brain tissues. MRI is the greatest tool for detecting such alterations in clinically relevant timeframes (Raja et al. 2018). There are two groups of CA-based approaches, i.e., Dynamic susceptibility contrast-enhanced MRI (DSC-MRI) and Dynamic contrast-enhanced MRI (DCE-MRI). For cerebral blood flow (CBF) and cerebral blood volume (CBV) measurements, DSC-MRI is employed, while DCE-MRI is used for permeability measures (Sourbron et al. 2009).

An in vivo study was conducted in rats to analyze the capillary permeability and perfusion in gliomas at different tumor stages using DCE and DSC MRI. CBV and capillary permeability changed temporally and spatially in all rats as the tumor progressed (Huhndorf et al. 2016). In human intracerebral hemorrhage, DCE MRI can evaluate and quantify BBB leaking in the brain region immediately surrounding the hematoma (Aksoy et al. 2013). This study determined that bigger hematomas, as well as hematomas in lobar sites, had more BBB leakage at 1 week, which is related to larger edema volumes. Moreover, Backes' group used the MRI approach to investigate BBB impairment in Alzheimer's disease (AD). For example, they found a link between BBB disruption and early AD in a group of 33 elderly, 9 of whom had mild cognitive impairment and 7 of whom had been diagnosed with AD. Indeed, individuals with Alzheimer's disease had more BBB leakage within the brain, which aggravated with cognitive impairment (Chagnot et al. 2021; Haar et al.

2016). Additionally, the same authors found an overall decline in cerebral blood flow in the gray matter of early AD patients in a follow-up investigation, which was connected with greater BBB leak in a cohort of 14 mild cognitive impairment/AD and 16 controls (Chagnot et al. 2021; van de Haar et al. 2016).

In order to increase the accuracy of MRI, quantitative ultrashort time-to-echo contrast-enhanced (QUTE-CE) MRI can be utilized as compared to gadolinium-based contrast agent (Gharagouzloo et al. 2015). QUTE-CE MRI combines ultra-short-time-to-echo sequences with ferumoxytol, superparamagnetic iron oxide nanoparticles (SPIONs), to yield positive contrast angiograms with minimal quantification error (Gharagouzloo et al. 2015, 2017). A study was done in BBZDR/Wor rat (type 2 diabetes rats' model) using QUTE-CE MRI to evaluate the BBB permeability with the hypothesis that small vessel dysfunction is a contributing factor to diabetes neuropathology (Qiao et al. 2020). The global capillary pathology in these rats' model with no insulin administration showed approximately more than 84% of the brain demonstrating a substantial escalation in BBB penetrability compared to controls. This study by QUTE-CE MRI indicated that diabetic neuro-pathologies such as encephalopathy and dementia are likely caused by an increase in BBB permeability (Qiao et al. 2020). A recent study developed a novel claudin 3/−4-sensitive xenon Hyper-CEST biosensor based on C-terminal claudin-binding domain of cCPE for the detection of claudins in tumors. The study incubated non-transfected HEK293 cells and HEK293 cells expressing claudin-4-FLAG with the cCPE-based biosensor and utilized xenon Hyper-CEST MRI for the detection of claudin-4 expressing cells (Piontek et al. 2017). Another recent study developed a nanoparticle (C1C2-NP) that specifically targeted regions with increased claudin-1 expression (Bony et al. 2021). Using dynamic contrast MRI, the study showed increased accumulation of C1C2-NP in the brain endothelial cells of mice with increased claudin-1 expression (Bony et al. 2021). Another study utilized the molecular-targeted MR imaging approach to visualize claudin-2 expression, which is a known urothelial permeability marker, in a lipopolysaccharide (LPS)-exposed transgenic mouse model for interstitial cystitis (URO-MCP-1) (Smith et al. 2020). Claudin-5 is a dominant TJ protein in the brain endothelial cells that is involved in the organization of TJs and in the modulation of BBB permeability through its interaction with the claudin-5 extracellular loops of adjacent endothelial cells (Ma et al. 2017). Similarly, a study used contrast-enhanced MRI for the in vivo measurements of BBB permeability in claudin-5 knockdown mice (Greene et al. 2018). The study reported that increased contrast in the acquired images were a result of Gd (contrast agent) extravasation in the claudin-5 knockdown mice (Greene et al. 2018) (Fig. 4.5).

4.5 Ultrasound

Ultrasound technology is an inexpensive, portable, and versatile imaging modality that has improved over time (Sastry et al. 2017; Feleppa et al. 2011). The related imaging techniques like 3-dimensional ultrasound, contrast-enhanced ultrasound,

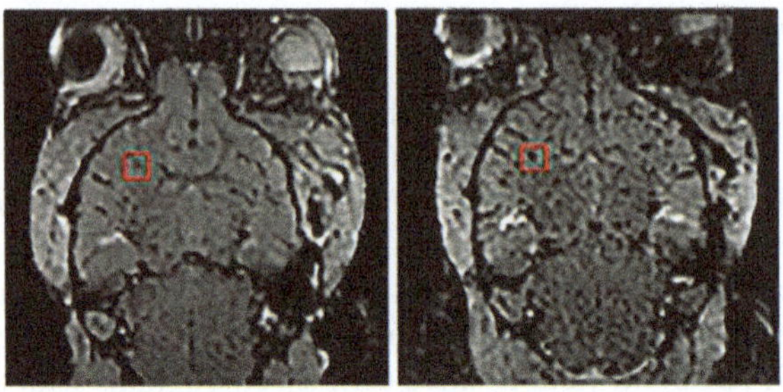

Fig. 4.5 Measurement of blood-brain barrier (BBB) permeability using contrast-enhanced MRI. Increased contrast can be observed in the brains of claudin-5 knockdown mice (right) compared to control mice (left). Reproduced with permission from Molecular Psychiatry and Greene et al. (2018)

high-intensity focused ultrasound (HIFUS), and ultrasonic elastography assure broader clinical usage (Sastry et al. 2017). Ultrasound can be broadly divided into two categories: diagnostic and therapeutic (ter Haar 2007). Diagnostic ultrasound is a type of imaging technique that employs sound waves to generate an image of organs, and other structures inside the body. Therapeutic ultrasound although also employs sound waves but it does not create images rather its goal is to interact with body tissues in such a way to either modify or destroy it. This section will focus on the use of ultrasound as a therapeutic technique.

The BBB not only prevents circulating molecules from entering the brain, but it also prevents the majority of medicinal and imaging substances from entering. One of the ways to circumvent the BBB and facilitate medication transfer is injecting an inert hypertonic solution intracarotidally, causing the endothelial cells to shrink and the interendothelial clefts to dilate (Sheikov et al. 2008). However, the immunoelectron microscopic analysis revealed alterations in the location and expression of tight junction proteins in the osmotically dilated clefts (Sheikov et al. 2008). Hence, this method has several shortcomings. To overcome this, the focused ultrasound (FUS) approach can be used that was first introduced by Lynn et al. in 1942 (Lynn et al. 1942), and then Fry et al. (1958). FUS alone can impact neuronal activity; however, combining FUS with circulating microbubbles (MBs) to open BBB can produce neurological and behavioral consequences, along with the effects that may last considerably longer than FUS alone (Munoz et al. 2018). MBs have gas-filled cores and can carry therapeutic agents such as drugs. Once MBs reach their targeted region, ultrasounds can then be utilized to rupture them resulting in site-specific delivery of therapeutic agents (Tsutsui et al. 2004). This makes MBs beneficial combined with ultrasound to be used in the administration of diagnostic and therapeutic drugs locally into the brain during the timeframe of BBB disruption that lasts several hours. Until now, FUS combined with MBs has been used to deliver various substances including genes (Lin et al. 2016), molecular imaging agents (Fan et al. 2016), chemotherapeutic agents (Fan et al. 2015), nanoparticles

(Zhao et al. 2016), and even cells (Alkins et al. 2016) to targeted tissues or organs (Wang et al. 2019).

One of the studies explored the effects of the MBs dosage on mice brain and found that with increasing MBs dosage the amount of BBB damage increases (Zhao et al. 2017). At specific MBs dosage and ultrasound exposure, considerably wider amount of BBB breakdown resulted with minimum tissue injury. Immunohistofluorescence assay showed significantly decreased positive cells of ZO-1, occludin, and claudin-5. Similarly, downregulation was seen in these tight junction proteins upon western blot analysis (Zhao et al. 2017). The decrease in intensity and expression of these proteins caused the disturbances in the BBB leading to increased BBB permeability. Hence, this study inferred that FUS administered with MBs can be used in the detection and treatment of brain illnesses by boosting BBB permeability to enhance delivery of therapeutic substances in the brain. Though focused ultrasound interaction with MBs allows the opening of BBB, however, the ultrasonic parameters must be fine-tuned: no opening will occur if the acoustic pressure is too low whereas bleeding will occur if pressure is too high. Recently a study was conducted to analyze if BBB permeability can be boosted by combining FUS^{+MB} with a second modality, allowing for lower acoustic pressures to be employed in a clinical setting (Chen et al. 2022). The authors of this study compared the effects of single, FUS^{+MB}, with combination therapy, FUS+MBs +cCPEm (a truncated enterotoxin form, previously been shown to reduce the BBB) (Chen et al. 2022). The MDCK II cell line (eGFP-hCldn5-MDCK II) pretreatment with cCPEm before FUS+MB treatment resulted in more barrier opening than either FUS+MB or cCPEm alone (Chen et al. 2022) (Fig. 4.6). This finding implied that pre-incubation with therapeutically appropriate tight junction protein binders might be used as a general method for making ultrasound-mediated BBB opening safer and more successful. Besides, a study investigated the effects of high-intensity focused ultrasound (HIFUS) in patients with prostate cancer and found that HIFU can be used as an alternative therapeutic approach (Lee et al. 2006).

4.6 Conclusion

Molecular imaging techniques can be effectively employed to detect and characterize an anomaly in tight junctions, as these imaging tools improve our understanding of intracellular communications. TJs are highly dysregulated in cancer and other ailments. Therefore, they are considered fundamental prognostic biomarkers to understand the stage and nature of such diseases. Molecular imaging techniques can accelerate drug development by early identification of abnormalities in TJs, eventually providing better therapeutic treatments. The response of these treatments could also be predicted through different molecular imaging tools. Further studies are required to determine the clinical utility of the above-mentioned molecular imaging technologies.

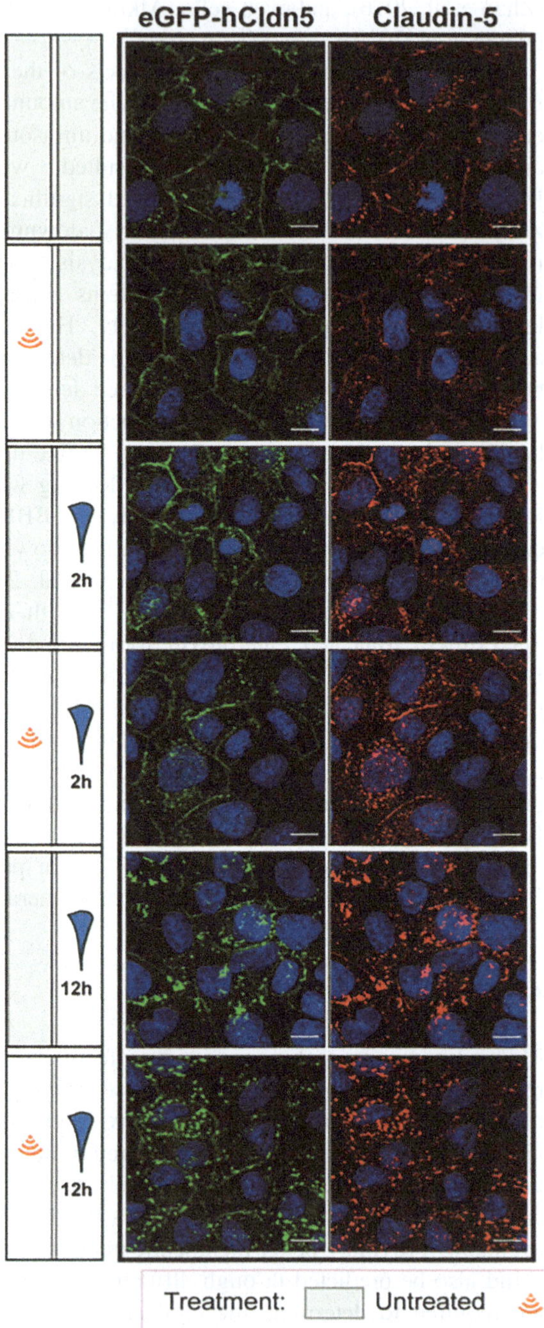

Fig. 4.6 Effect of glutathione S-transferase (GST) cCPEm fusion protein and focused ultrasound (FUS) intravenously injected microbubbles (FUS^{+MB}) on claudin-5 expression alone and in combination in MDCK II cells. Combination treatment with GST-cCPEm and FUS^{+MB} resulted in increased translocation of claudin-5 from membrane into the cytosol, destabilizing the tight junctions (TJs). Reproduced with permission from Theranostics and Chen et al. (2022)

References

Aksoy D et al (2013) Magnetic resonance imaging profile of blood-brain barrier injury in patients with acute intracerebral hemorrhage. J Am Heart Assoc 2(3):e000161

Alkins R et al (2016) Early treatment of HER2-amplified brain tumors with targeted NK-92 cells and focused ultrasound improves survival. Neuro-Oncology 18(7):974–981

Amoo M et al (2022) Permeability of the blood-brain barrier after traumatic brain injury: radiological considerations. J Neurotrauma 39(1-2):20–34

Baguña Torres J et al (2020) Radiolabeled cCPE peptides for SPECT imaging of claudin-4 overexpression in pancreatic cancer. J Nucl Med 20:243113

Bartosova M et al (2021) An experimental workflow for studying barrier integrity, permeability, and tight junction composition and localization in a single endothelial cell monolayer: proof of concept. Int J Mol Sci 22(15):8178

Bednarczyk J, Lukasiuk K (2011) Tight junctions in neurological diseases. Acta Neurobiol Exp 71(4):393–408

Bony BA et al (2021) Claudin-1-targeted nanoparticles for delivery to aging-induced alterations in the blood-brain barrier. ACS Nano 15:18520–18531

Bybel B et al (2008) SPECT/CT imaging: clinical utility of an emerging technology. Radiographics 28(4):1097–1113

Chagnot A, Barnes SR, Montagne A (2021) Magnetic resonance imaging of blood-brain barrier permeability in dementia. Neuroscience 474:14–29

Chen L et al (2022) Claudin-5 binder enhances focused ultrasound-mediated opening in an in vitro blood-brain barrier model. Theranostics 12(5):1952–1970

Cheng C et al (2015) Gene expression profiling of colon cancer by correlation with 18F-FDG kinetics as measured by dynamic positron emission tomography-computed tomography (dPET-CT): dependency on cadherin-related genes and hypoxia. J Nucl Med 56(3):1341

Fan CH et al (2015) Drug-loaded bubbles with matched focused ultrasound excitation for concurrent blood-brain barrier opening and brain-tumor drug delivery. Acta Biomater 15:89–101

Fan CH et al (2016) Ultrasound/magnetic targeting with SPIO-DOX-microbubble complex for image-guided drug delivery in brain tumors. Theranostics 6(10):1542–1556

Farquhar MG, Palade GE (1963) Junctional complexes in various epithelia. J Cell Biol 17(2): 375–412

Feleppa EJ et al (2011) Quantitative ultrasound in cancer imaging. Semin Oncol 38(1):136–150

Foss CA et al (2007) Radiolabeled anti-claudin 4 and anti-prostate stem cell antigen: initial imaging in experimental models of pancreatic cancer. Mol Imaging 6(2):131–139

Fry FJ, Ades HW, Fry WJ (1958) Production of reversible changes in the central nervous system by ultrasound. Science 127(3289):83–84

Furuse M et al (1998) Claudin-1 and -2: novel integral membrane proteins localizing at tight junctions with no sequence similarity to occludin. J Cell Biol 141(7):1539–1550

Gharagouzloo CA, McMahon PN, Sridhar S (2015) Quantitative contrast-enhanced MRI with superparamagnetic nanoparticles using ultrashort time-to-echo pulse sequences. Magn Reson Med 74(2):431–441

Gharagouzloo CA et al (2017) Quantitative vascular neuroimaging of the rat brain using superparamagnetic nanoparticles: new insights on vascular organization and brain function. NeuroImage 163:24–33

Greene C et al (2018) Dose-dependent expression of claudin-5 is a modifying factor in schizophrenia. Mol Psychiatry 23(11):2156–2166

Gumbiner B, Lowenkopf T, Apatira D (1991) Identification of a 160-kDa polypeptide that binds to the tight junction protein ZO-1. Proc Natl Acad Sci U S A 88(8):3460–3464

Haar HJ et al (2016) Blood-brain barrier leakage in patients with early Alzheimer disease. Radiology 281(2):527–535

Haskins J et al (1998) ZO-3, a novel member of the MAGUK protein family found at the tight junction, interacts with ZO-1 and occludin. J Cell Biol 141(1):199–208

Higashi K et al (2017) High FDG uptake on PET is associated with negative cell-to-cell adhesion molecule E-cadherin expression in lung adenocarcinoma. Ann Nucl Med 31(8):590–595

Huhndorf M et al (2016) Alterations of the blood-brain barrier and regional perfusion in tumor development: MRI insights from a rat C6 glioma model. PLoS ONE 11(12):e0168174

Hwang TL et al (2010) Claudin-4 expression is associated with tumor invasion, MMP-2 and MMP-9 expression in gastric cancer. Exp Ther Med 1(5):789–797

Hwang TL et al (2014) Claudin-4 expression in gastric cancer cells enhances the invasion and is associated with the increased level of matrix metalloproteinase-2 and -9 expression. Oncol Lett 8(3):1367–1371

Jesaitis LA, Goodenough DA (1994) Molecular characterization and tissue distribution of ZO-2, a tight junction protein homologous to ZO-1 and the Drosophila discs-large tumor suppressor protein. J Cell Biol 124(6):949–961

Jiwa LS et al (2014) Upregulation of Claudin-4, CAIX and GLUT-1 in distant breast cancer metastases. BMC Cancer 14(1):864

Johnson LG (2005) Applications of imaging techniques to studies of epithelial tight junctions. Adv Drug Deliv Rev 57(1):111–121

Khan S et al (2020) High-resolution positron emission microscopy of patient-derived tumor organoids. bioRxiv, p 2020.07.28.220343

Kim WS et al (2022) High expression of claudin-4 is associated with synchronous tumors in patients with early gastric cancer. J Clin Med 11(12):3550

Knight RA et al (2009) Estimating blood and brain concentrations and blood-to-brain influx by magnetic resonance imaging with step-down infusion of Gd-DTPA in focal transient cerebral ischemia and confirmation by quantitative autoradiography with Gd-[(14)C]DTPA. J Cereb Blood Flow Metab 29(5):1048–1058

Lee HM, Hong JH, Choi HY (2006) High-intensity focused ultrasound therapy for clinically localized prostate cancer. Prostate Cancer Prostatic Dis 9(4):439–443

Li X-R et al (2021) PET-CT radiomics by integrating primary tumor and peritumoral areas predicts E-cadherin expression and correlates with pelvic lymph node metastasis in early-stage cervical cancer. Eur Radiol 31(8):5967–5979

Lin CY et al (2016) Non-invasive, neuron-specific gene therapy by focused ultrasound-induced blood-brain barrier opening in Parkinson's disease mouse model. J Control Release 235:72–81

Liu C et al (2016) Development of PET probes for molecular imaging of tight junction membrane proteins, claudin-4, in ovarian cancer. J Nucl Med 57(2):530

Lochhead JJ et al (2020) Structure, function, and regulation of the blood-brain barrier tight junction in central nervous system disorders. Front Physiol 11:914

Lynn JG et al (1942) A new method for the generation and use of focused ultrasound in experimental biology. J Gen Physiol 26(2):179–193

Ma X et al (2015) Claudin-4 controls the proliferation, apoptosis, migration and in vivo growth of MCF-7 breast cancer cells. Oncol Rep 34(2):681–690

Ma SC et al (2017) Claudin-5 regulates blood-brain barrier permeability by modifying brain microvascular endothelial cell proliferation, migration, and adhesion to prevent lung cancer metastasis. CNS Neurosci Ther 23(12):947–960

Madsen MT (2007) Recent advances in SPECT imaging. J Nucl Med 48(4):661–673

Martin TA (2014) The role of tight junctions in cancer metastasis. Semin Cell Dev Biol 36:224–231

Mosley M et al (2014) Abstract 4929: Radiolabeled cCPE for molecular imaging of tight junction changes during breast oncogenesis. Cancer Res 74(19):4929

Mosley M et al (2015) Claudin-4 SPECT imaging allows detection of aplastic lesions in a mouse model of breast cancer. J Nucl Med 56(5):745

Munoz F et al (2018) Modulation of brain function and behavior by focused ultrasound. Curr Behav Neurosci Rep 5(2):153–164

Piontek J et al (2011) Elucidating the principles of the molecular organization of heteropolymeric tight junction strands. Cell Mol Life Sci 68(23):3903–3918

Piontek A et al (2017) A cCPE-based xenon biosensor for magnetic resonance imaging of claudin-expressing cells. Ann N Y Acad Sci 1397(1):195–208

Qiao J et al (2020) Evaluating blood–brain barrier permeability in a rat model of type 2 diabetes. J Transl Med 18(1):256

Raja R, Rosenberg GA, Caprihan A (2018) MRI measurements of blood-brain barrier function in dementia: a review of recent studies. Neuropharmacology 134(Pt B):259–271

Raleigh DR et al (2010) Tight junction-associated MARVEL proteins marveld3, tricellulin, and occludin have distinct but overlapping functions. Mol Biol Cell 21(7):1200–1213

Sastry R et al (2017) Applications of ultrasound in the resection of brain tumors. J Neuroimaging 27(1):5–15

Schillaci O (2004) Functional–anatomical image fusion in neuroendocrine tumors. Cancer Biother Radiopharm 19(1):129–134

Schillaci O (2005) Hybrid SPECT/CT: a new era for SPECT imaging? Eur J Nucl Med Mol Imaging 32(5):521–524

Sheikov N et al (2008) Effect of focused ultrasound applied with an ultrasound contrast agent on the tight junctional integrity of the brain microvascular endothelium. Ultrasound Med Biol 34(7): 1093–1104

Shen L et al (2011) Tight junction pore and leak pathways: a dynamic duo. Annu Rev Physiol 73: 283–309

Slifer ZM, Blikslager AT (2020) The integral role of tight junction proteins in the repair of injured intestinal epithelium. Int J Mol Sci 21(3):972

Smith N et al (2020) In vivo and ex vivo assessment of bladder hyper-permeability and using molecular targeted magnetic resonance imaging to detect claudin-2 in a mouse model for interstitial cystitis. PLoS One 15(10):e0239282

Sourbron S et al (2009) Quantification of cerebral blood flow, cerebral blood volume, and blood-brain-barrier leakage with DCE-MRI. Magn Reson Med 62(1):205–217

Tabariès S, Siegel PM (2017) The role of claudins in cancer metastasis. Oncogene 36(9):1176–1190

ter Haar G (2007) Therapeutic applications of ultrasound. Prog Biophys Mol Biol 93(1):111–129

Torigian DA et al (2007) Functional imaging of cancer with emphasis on molecular techniques. CA Cancer J Clin 57(4):206–224

Torres JB et al (2018) Imaging of claudin-4 in pancreatic ductal adenocarcinoma using a radiolabelled anti-claudin-4 monoclonal antibody. Mol Imaging Biol 20(2):292–299

Tsukita S, Tanaka H, Tamura A (2019) The claudins: from tight junctions to biological systems. Trends Biochem Sci 44(2):141–152

Tsutsui JM, Xie F, Porter RT (2004) The use of microbubbles to target drug delivery. Cardiovasc Ultrasound 2:23–23

Ueda J et al (2007) Heterogeneous expression of claudin-4 in human colorectal cancer: decreased claudin-4 expression at the invasive front correlates cancer invasion and metastasis. Pathobiology 74(1):32–41

van de Haar HJ et al (2016) Neurovascular unit impairment in early Alzheimer's disease measured with magnetic resonance imaging. Neurobiol Aging 45:190–196

Vander Heiden MG, Cantley LC, Thompson CBJ (2009) Understanding the Warburg effect: the metabolic requirements of cell proliferation. Science 324(5930):1029–1033

Vanslembrouck B et al (2022) Microscopic visualization of cell-cell adhesion complexes at micro and nanoscale. Front Cell Dev Biol 10:819534

Verin AD, Bogatcheva NV (2006) Cytoskeletal proteins. In: Laurent GJ, Shapiro SD (eds) Encyclopedia of respiratory medicine. Academic Press, Oxford, pp 615–622

Wang Y et al (2016) Rhubarb attenuates blood-brain barrier disruption via increased zonula occludens-1 expression in a rat model of intracerebral hemorrhage. Exp Ther Med 12(1): 250–256

Wang B et al (2019) A novel WARS mutation (p.Asp314Gly) identified in a Chinese distal hereditary motor neuropathy family. Clin Genet 96(2):176–182

Weissleder R, Mahmood U (2001) Molecular imaging. Radiology 219(2):316–333

Yasuda S et al (2000) Application of positron emission tomography imaging to cancer screening. Br J Cancer 83(12):1607–1611
Zhao YZ et al (2016) Glioma-targeted therapy using cilengitide nanoparticles combined with UTMD enhanced delivery. J Control Release 224:112–125
Zhao B et al (2017) Blood-brain barrier disruption induced by diagnostic ultrasound combined with microbubbles in mice. Oncotarget 9(4):4897–4914
Zhao C et al (2022) Targeting claudin 18.2 using a highly specific antibody enables cancer diagnosis and guided surgery. Mol Pharm 19:3530–3541
Zihni C et al (2016) Tight junctions: from simple barriers to multifunctional molecular gates. Nat Rev Mol Cell Biol 17(9):564–580

Tight Junction Proteins as Emerging Drug Targets: Expanding the Horizons from Inflammation to Cancer

5

Sireesha V. Garimella, Rahul Roy, Siri Chandana Gampa, and Santhi Latha Pandrangi

Abstract

Tight junctions (TJ), located at the most topical part of the cells, are multi-protein complexes that form a formidable barrier governing the permeability of polarized epithelial and endothelial cells. They principally function as an intercellular barrier in mediating adhesion and maintaining cellular polarity. There is gathering evidence that dysregulated cell-cell adhesion is associated with numerous cancers and inflammatory diseases. A large number of cancers originate from epithelial cells, and several TJ proteins are seen dysregulated in the inflamed intestinal tract and malignant tumors. Irregularities in the normal functioning of the TJs might facilitate cancer initiation, progression, or metastasis, or might facilitate the progression of inflammatory diseases. Thus, TJs play a crucial role in regulating the influence on cell polarity, cell fate, and cell migration. This chapter outlines the paradigm of the dysregulated TJ proteins and their consequent effect in cell-cell interaction leading to inflammation and cancer, and how targeting of TJ proteins as drug candidates offers a promise in combating these diseases, providing platforms for the development of novel therapies.

S. V. Garimella (✉) · S. C. Gampa
Department of Biotechnology, GITAM Institute of Science, GITAM (Deemed to be University), Visakhapatnam, India
e-mail: sgarimel@gitam.edu

R. Roy
Centre for Biomedical Engineering, Indian Institute of Technology, New Delhi, India

S. L. Pandrangi
Department of Biochemistry and Bioinformatics, GITAM Institute of Science, GITAM (Deemed to be University), Visakhapatnam, India

A. A. Bhat et al. (eds.), *Tight Junctions in Inflammation and Cancer*,
https://doi.org/10.1007/978-981-99-2415-8_5

Keywords

Cancer · Inflammation · Tight junction (TJ) · Junctional adhesion molecule (JAM) · Claudin · Zonula occludens (ZO) · Tumorigenesis · Adhesion · Metastasis

5.1 Introduction

5.1.1 Tight Junctions: Type, Classification, and Structure

Every living cell has specialized interacting multi-protein complexes called the cell junctions that occur at cell-to-cell and cell-matrix contacts of all animal tissues. Four major cell junctions exist: tight junctions, adherens junctions, desmosomes, and gap junctions (Itoh and Bissell 2003). Tight junctions (TJ), also known as occludin junctions or Zonula occludens (ZO), comprise two proteins called occludins and claudins and are mainly found in epithelial cells. Adhering junctions, also known as Zonula adherens (ZA), are protein complexes comprising cadherin receptors and cytoplasmic adapter proteins called catenins. These junctions have cytoplasmic faces linked to the actin cytoskeleton that initiate cell-cell contacts, provide strong mechanical attachments, and support the concentration of scaffolding and signaling molecules (Hartsock and Nelson 2008). Another type of cell junction is the Macula adherens (MA) or desmosomes. Desmosomes provide strong adhesion between the cells with the help of cadherin family proteins, desmogleins, and desmocollins. They protect the tissues from mechanical stress and maintain tissue integrity (Garrod and Chidgey 2008; Kowalczyk and Green 2013). Gap junctions, also known as nexus or Macula communicans (MC), are a cluster of intercellular channels composed of transmembrane proteins called connexins and allow direct cell-cell transfer of ions, small molecules, and second messengers between adjacent cells cytoplasm (Mese et al. 2007; Goodenough and Paul 2009).

Out of the four major types of cell junctions mentioned above, the TJs are of particular interest as these originate from specialized differentiation of epithelial and endothelial cell membranes and play a very crucial role in adhesion and interaction of diverse cell types. In epithelial cells, tight junctions are asymmetrically distributed at the apical side of the lateral membrane that forms the apical junctional complex encircling the apex of the cells and forms a border between the apical and basolateral membranes (Fig. 5.1). TJs are multi-protein junctional complexes involved in a multitude of functions by serving as a barrier between the intercellular spaces and partitioning the apical and basolateral fluid compartments of epithelia and endothelia. Depending on the protein composition, they regulate epithelial and endothelial cell permeability and, in addition, help in maintaining cell polarity (Nelson 2003).

The TJs are known to form and regulate the paracellular barrier between epithelial and endothelial cell sheets. Apart from barrier function, they also play a role in cell-cell adhesion and an intramembrane diffusion fence (Tsukita et al. 2008; Martin et al.

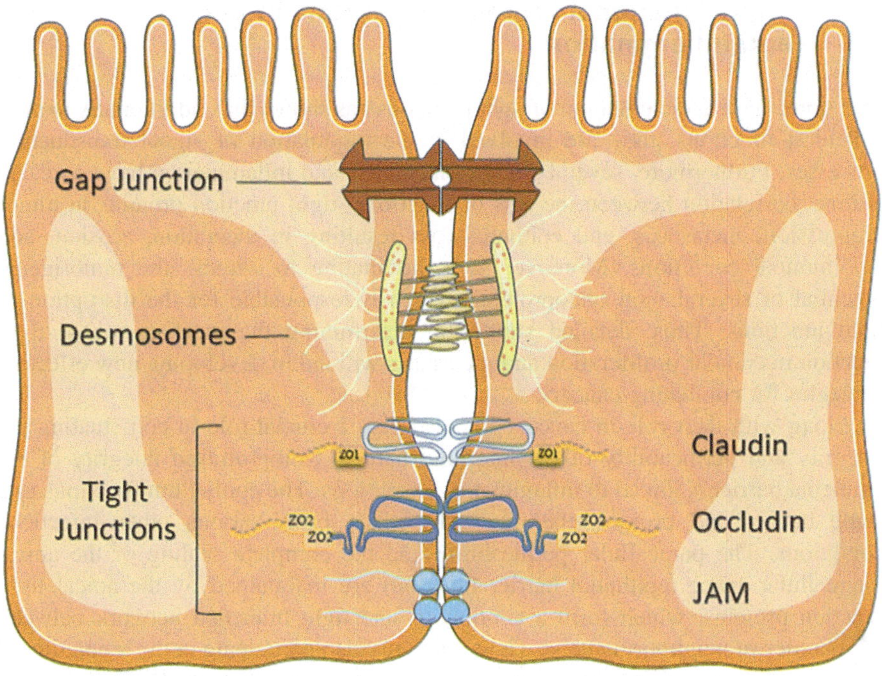

Fig. 5.1 Junctional complexes that facilitate cross-talk between two adjacent cells

2002).The TJs fence function restricts the intermixing of molecules between the apical and lateral membrane and helps keep a discrete state. This function distinguishes between the structural and functional domains, thus maintaining cell polarity, and is deeply involved in cancer cell biology in terms of loss of cell polarity (Gasbarrini and Montalto 1999; Hollander 1999; Teshima et al. 2012).

Moreover, the barrier function of TJs has been shown to play a significant role in differentiation, regulation of cell growth, and proliferation. Tight junctions maintain high resistance barriers, unlike other cell junctions. Based on the size and charge, tight junctions act as the primary regulators of fluid and solute flux; the function is termed "permselectivity." In addition, tight junctions regulate tension, play a role in cell signaling by transmitting signals into nuclei, and modulate gene expression.

The central theme of this chapter is to discuss the altered modulation of cell-cell junctions, particularly tight junctions, resulting in the development of cancer phenotypes, inflammation, and the potential of tight junction proteins to be used as drug targets in cancer and inflammation.

5.2 Tight Junctional Proteins: Their Role in Cancer and Inflammation

The molecular characterization of tight junctions has shown that, additionally to their structural functions, they are involved in the modulation of signal transduction processes. Furthermore, several studies on cancers and inflammation have revealed a direct correlation between the loss of functional tight junction proteins in tumor progression, metastasis, and epithelial loss resulting in ulceration, erosion, and autoimmune conditions. Moreover, it is fascinating to witness the tumorigenic potential of several viral oncoproteins, which is responsible for the disruption of tight junctions. Thus, detailed studies in determining the mechanisms of tight junction in cellular proliferation and metastasis will aid in developing new effective strategies for combating cancers.

Along with their role in cancers TJs also play a crucial role in perpetuating the integrity and permeability of the intestinal barrier. Compromised integrity of the epithelial barrier is linked to inflammatory conditions. The epithelium functions as a static barrier and protects the mucosal surface from various other infectious conditions. The paracellular permeability and the complete sealing of the apical intercellular space (epithelial barrier function) are maintained by the apical tight junction proteins, which form a continuous and tight branched network between membranes of neighboring cells. Tight junctions also participate in the regulation of gene expression required for cell proliferation and differentiation. Thus, alterations in the tight junction proteins are seen in inflammatory bowel diseases and provide a good target for developing novel therapeutics (Matter et al. 2005).

The tight junctional complex consists of several proteins, including a set of transmembrane proteins and the cytoplasmic scaffold proteins that connect the transmembrane proteins to the cytoskeleton of the cell. The transmembrane proteins include occludins, claudins, and junctional adhesion molecules (JAMs). The plaque proteins comprise the zona occludins, PDZ domains, MAGUK, the Crumbs/PALS1/PATJ complex, and the PAR-3/PAR-6/aPKC complex. Each of these proteins are discussed in detail below.

5.2.1 Occludin

Occludins share a general architecture as tetraspan transmembrane proteins of claudins. This first discovered novel integral membrane protein was identified as the tight junction component of hepatic junctional fractions and had four transmembranes MARVEL (MAL and related proteins for vesicle trafficking and membrane link) domains of ~65 kDa (Weber and Turner 2007; Clayburgh et al. 2004; Hollander 1988). Despite it being speculated to be indispensable for tight junction structural assembly, further knockouts of mice for occludin suggested that it was not necessarily required for tight junction strand formation (Munkholm et al. 1994). Furthermore, it was seen that such knockouts resulted in histological abnormalities like chronic inflammation and hyperplasia of gastric epithelium with loss of parietal chief cells, dwarf-born mice, brain calcifications, osteopenia, defective epithelial barrier, and testicular atrophy (Munkholm et al. 1994; Bruewer et al. 2006). Furthermore, the role of occludin was reported to be in maintaining the cell polarity, and elements of the hippo pathway co-localize with occludin, which may be involved in the modulation of the proliferation of the pancreatic epithelial cells.

Studies in human cutaneous squamous cell carcinoma (SCC) have shown that occludin downregulation has cellular characteristics for tumorigenesis resulting in decreased epithelial cell-cell adhesion and reduced susceptibility to apoptosis induced by UVB or TRAIL (TNF-related apoptosis-inducing ligand), altered Ca^{2+} homeostasis which may contribute to alterations of cell-cell adhesion and differentiation (Hollander et al. 1986). Moreover, occludin suppresses cellular invasiveness and motility, consequently abrogating metastatic properties of cancer cells and suggesting occludin as a likely candidate for a tumor suppressor gene and target for therapeutic intervention in certain cancer types (Haynes et al. 2005; Sawada 2013; Shin et al. 2006; Anderson and Van Itallie 2009). The overexpression of occludin promoted anoikis in breast carcinoma cells while endogenous occludin re-expression correlated with downregulation of apoptosis-inhibitory genes and upregulation of apoptosis-inducing genes. TUNEL assays revealed that HeLa cells constitutively overexpressing wild-type occludin exhibited increased sensitivity to oxidative stress-induced apoptosis (Brennan et al. 2010).

Similar to occludin, another novel integral protein identified was tricellulin; both contain tetraspaning MARVEL domains of ~70 kDa, and it is concentrated at tricellular tight junctions, and it plays a role in the formation of bicellular and tricellular tight junctions (Boireau et al. 2007; Brennan et al. 2010). In contrast to occludin knockout, tricellulin suppression led to the compromised epithelial barrier, and disorganized tricellular contacts and bicellular tight junctions were seen, indicating the vital function of tricellulin for epithelial barrier formation (Brennan et al. 2010).

5.2.2 Claudins

Among all the tight junction proteins identified, the claudin family of transmembrane proteins has emerged as the most critical for defining tight junction selectivity. The four transmembrane domains of ~23 kDa of claudins are necessary for the tight junction formation, suggesting that it constitutes the tight junctions' structural element and molecular backbone (Fedwick et al. 2005; Fink et al. 2006; Grone et al. 2007). Studies have reported that functional analysis of claudins has a pivotal role in the regulation of cell proliferation, and thus providing therapeutic targets for developing anticancer strategies (Tsukita et al. 2008). Moreover, studies have shown that claudins, in particular, are critical components of tight junctions, and these have been implicated as key modulators of carcinogenesis and metastasis in numerous cancer types. The expression patterns and the claudin type differ depending on the type of cancer, which plays a key role in tumor invasion and aggressiveness, thus delimiting the patient's prognosis (Kohno et al. 2006; Tzelepi et al. 2008; Ip et al. 2007; Paschoud et al. 2007; Raskov et al. 2017).

Dysregulated expression of claudin-1, claudin-2, claudin-3, claudin-4, and claudin-7 was reported in breast cancer growth and malignancy via the control of cell proliferation, migration, metastasis, and apoptosis (Bhat et al. 2018; Salvador et al. 2016; Forster 2008; Landy et al. 2016). Likewise, in pancreatic cancers, there has been an observation of the expression of claudin-1 and the activation of mitogen-activated protein kinases (MAPK) signaling. On the other hand, decreased expression of claudins-7 has been reported in head and neck cancers, metastatic breast cancer, and invasive ductal adenocarcinomas (Landy et al. 2016; Furuse et al. 1993). Moreover, there has been a correlation between altered expression of claudin-4 and human bladder tumors, as reported by Furuse (Furuse et al. 1993). The gene expression analysis encoding tight junction proteins revealed the overexpression of claudin-1 and -12 in colorectal cancer and claudin-8 downregulation in tumor tissue on RNA level. Moreover, protein quantification and immunohistochemistry studies confirmed the overexpression of claudin-1 in tumor tissues (Grone et al. 2007). Claudin-4, -7, and -10 were identified as biphasic tumor-related claudins in synovial sarcoma and are present in the epithelial component of biphasic tumors (Kohno et al. 2006). In vitro assays with CLDN-10 overexpression and small interfering RNA-mediated knockdown transfectants developed hepatocellular carcinoma (Ip et al. 2007). Lower expression of CLDN-1 and -5 showed higher prostate-specific antigen in prostate cancers (Raskov et al. 2017).

In the GI tract, claudins are mainly involved in building paracellular selective channels, acting as signaling proteins mediating cell behaviors. Moreover, their dysfunction has been indicated in the pathogenesis of inflammatory bowel disease, epithelial permeation disorders, and multiple intestinal diseases (Tsukita and Tsukita 1989). These findings of claudins dysregulated expression, cellular distribution, and its association in various cancers and inflammatory diseases shed new light on further etiologic research and may serve as important markers for better prognosis and for developing potential therapeutics in clinical treatments.

5.2.3 Junctional Adhesion Molecules

Junctional adhesion molecules (JAMs) belong to the third class of integral membrane proteins, including the members of the immunoglobulin superfamily (Tsukita and Furuse 1999; Saitou et al. 2000; Schulzke et al. 2005; Rachow et al. 2013; Osanai et al. 2006). JAMs are ~43 kDa glycosylated proteins, and they are characterized by two extracellular V-type Ig domains, a transmembrane domain, and a short intracellular COOH with a PDZ motif (Schulzke et al. 2005). These single transmembrane proteins are tight junction strand-constitutive proteins involved in regulating cell interactions in the immune system, immune cell transmigration, and the formation of tight junctions in epithelial and endothelial cells (Orban et al. 2008; Tobioka et al. 2004). They also function in the regulation of paracellular permeability to circulating cells and solutes (Osanai et al. 2006). JAM is present in the transmembrane component of the blood tissue barriers, and they are involved in numerous cellular functions such as leukocyte migration and angiogenesis; they also are responsible for pathological conditions like hypertension and tumorigenesis (Martin et al. 2010). In addition, they are also expressed in the leukocytes and platelets, due to which these proteins are not only associated with the permeability barrier function of the tight junctions but also are involved in inflammatory reactions (Martin et al. 2010). JAMA is a crucial extracellular adhesive molecule involved in controlling mucosal homeostasis by regulating the integrity and permeability of the intestinal barrier function. Moreover, altered structure and epithelial permeability are associated with inflammatory bowel disease (Sanchez-Pulido et al. 2002). Furthermore, the role of JAM family members in protumorigenic functions, including inhibition of apoptosis, promotion of proliferation, and epithelial to mesenchymal transition, has been implicated (Ikenouchi et al. 2005).

5.2.4 Zonula Occludens

Zona occludens (ZO) belong to a family of multidomain proteins known as membrane-associated guanylate kinase homologs (MAGUKs). In addition to the primary functioning of tight junctions as a paracellular barrier and as a fence, they also are involved in diverse processes such as cell polarity, cell proliferation, and differentiation. These functions require signal transduction from the plasma membrane, which is primarily carried out by the cytosolic tight junction plaque proteins (Tsukita and Furuse 2000; Furuse et al. 1998a). ZO are scaffold-forming intracellular plaque proteins of ~210–225 kDa located at the sub-membranous domain of tight junctions in epithelia and endothelia. They regulate the assembly of cellular junctions and is seen associated through the first PDZ domain of the carboxyl-terminal of claudins, second and third PDZs to JAM and its guanylate kinase-like domain (Guk) to the cytoplasmic carboxyl-terminal region of occludin (Furuse et al. 1998b; Tabaries and Siegel 2017).

There are two significant groups of ZOs: scaffolding proteins and signaling proteins. Peripherally associated scaffolding proteins like ZO-1, ZO-2, ZO-3, AF6, and cingulin play a role in forming links between the integral tight junction proteins and the actin cytoskeleton and other cytoplasmic proteins (Furuse et al. 1998a). The "signaling" proteins ZO-1-associated nucleic acid binding (ZONAB), RhoA, RalA, and Raf-1 function as adapters for integrating cytosolic molecules involved in cell signaling, barrier regulation, and gene transcription. Downregulation of ZO-1 expression and its failure to accumulate at the cellular junctions have been shown to correlate to breast cancer progression, increased invasiveness in breast cancer (Bhat et al. 2020), colorectal cancer (Osanai et al. 2017), liver cancer (Morin 2005), and gastrointestinal tumors (Singh and Dhawan 2015). ZO-1 is reported to be involved in tumor invasion associated with EMT (Ma et al. 2015). Furthermore, recent studies have shown that the systemic concentrations of ZO-1 are seen significantly elevated in hepatocellular carcinoma (HCC) that correlates with the inflammation, suggesting that inflammation promotes plasma ZO-1 concentration

(Kim et al. 2008). The immunostaining results of ZO-1 and ZO-2 proteins expression and distribution pattern were typical in normal seminiferous tubules compared to CIS tubules in the testis (Fink et al. 2006) (Table 5.1).

5.3 Targeting TJ Proteins for Therapeutic Application in Cancer and Inflammation

Tight junction components serve as promising therapeutic targets and are now emerging for the development of drugs to treat cancer and inflammatory diseases. With the potential to develop targeted therapies, and due to monoclonal antibodies' highly specific nature, several antibodies have been developed and are investigated against a few tight junctional proteins (Table 5.2). In epithelium-derived tumors, aberrant alterations result in the loss of both cellular polarity and epithelial integrity, leading to malignant transformation of epithelial cells, abnormal cell growth, gaining invasiveness, and metastatic potential (French et al. 2009). Among the TJ proteins, dysregulated claudins are seen in many cancers and inflammatory conditions (Ip et al. 2007). Presently, mAbs have been generated against Claudin-1, Claudin-2, Claudin-3, Claudin-4, Claudin-6, Claudin-11, and Claudin-18.2. Upregulated Claudin-3 and claudin-4 have been associated with malignant tumors of the breast (Arcangeli et al. 2012), ovarian (Ghislin et al. 2011), prostate (Langer et al. 2011), and pancreatic cancers (Santin et al. 2007). Anti-claudin-4 mAb (KM3934) has inhibited tumor growth in xenograft models of pancreatic and ovarian cancer (Weber et al. 2008). Similarly, attenuated tumor growth was seen in gastric and colon cancer with anti-claudin-4 mAb (5D12) (Hadj-Rabia et al. 2004). Also, anti-claudin-3 and -4 bispecific mAbs (KM3907 and 5A5) have shown antitumor effects in other cancers (Argaw et al. 2009; Prasad et al. 2005).

In addition, a similar association of claudin-1 and claudin-2 has been seen in colorectal and other cancers. Recent studies have investigated the antitumor properties of anti-claudin-1 and -2 mAbs. Treatment of colorectal cancer xenograft mice with an anti-claudin-1 mAb (6F6) suppressed the growth of the tumor, and similar results were obtained upon treatment with anti-claudin-2 mAb (1A2) in xenograft mice (Hadj-Rabia et al. 2004; Scharl et al. 2009). In colon cancer, claudin-1 and claudin-2 overexpression is attributed to gaining metastatic potential and increased invasiveness (Lauko et al. 2020)

Inhibitors of claudin-2 are promising novel therapeutic drugs for inflammatory bowel diseases, such as ulcerative colitis and Crohn's disease. These conditions are characterized by upregulation of claudin-2 expression resulting in chronic mucosal inflammation of the gastrointestinal tract, leading to the impairment of the intestinal barrier function (Sun et al. 2011; Takehara et al. 2009). Moreover, inflammatory cytokines, including tumor necrosis factor-α (TNF-α), have been shown to decrease the TJ integrity and increase claudin-2 expression. Thus, specific inhibition of claudin-2 and TNF-α could help in increasing the TJ integrity and thereby preventing intestinal inflammation. Recently, an anti-claudin-2 mAb (clone 1A2) was used to inhibit the claudin-2 expression in Caco-2 cells selectively, and

Table 5.1 Expression of TJ proteins in cancer and inflammation

	Claudin																Occludin	ZO		JAM		
Type	1	2	3	4	5	7	8	9	10	11	12	14	16	17	18	20		1	2	A	C	References
Breast	↑↓	↑↓	↑	↑		↓							↑			↑		↓		↑		Bhat et al. (2018), Forster (2008), Landy et al. (2016), Morohashi et al. (2007), Kominsky et al. (2003), Zhu et al. (2019)
Prostate		↓	↑	↑																		Morohashi et al. (2007), Ebnet (2008), Aurrand-Lions et al. (2001)
Lung	↑	↑	↑	↑		↑		↑	↑											↑	↑	Aurrand-Lions et al. (2001), Martin-Padura et al. (1998), Liu et al. (2000)
Pancreatic	↑		↑	↑		↑												↑				Ebnet (2008), Aurrand-Lions et al. (2001), Cohen et al. (2001)
Colorectal	↑↓	↑	↑	↓		↑↓	↓				↑						↑	↓				Ebnet et al. (2004), Bazzoni et al. (2000), Bazzoni (2011), Vetrano and Danese (2009), Lauko et al. (2020), Gonzalez-Mariscal et al. (2003), Schneeberger and Lynch (2004), Fanning et al. (1998)
Ovarian	↑		↑	↑		↑							↑								↑	Ebnet (2008), Aurrand-Lions et al. (2001), Schmidt et al. (2001), Hoover et al. (1998), Kaihara et al. (2003), Zhang et al. (2019), Kimura et al. (1997)
Colon	↑	↑↓	↓	↑↓		↑↓																Fanning et al. (1998), Polette et al. (2007), Ram et al. (2018), Soini (2005)

(continued)

Table 5.1 (continued)

Type	Claudin																Occludin	ZO		JAM		References
	1	2	3	4	5	7	8	9	10	11	12	14	16	17	18	20		1	2	A	C	
Gastric	↑		↑	↑		↑			↓	↑		↑		↓						↑		Ebnet (2008), Aurrand-Lions et al. (2001), Majer et al. (2016), Sauer et al. (2005), Rangel et al. (2003), Hewitt et al. (2006), Sharma et al. (2016)
Cervical	↑	↑		↑		↑																Zhang et al. (2013), Alikanoglu et al. (2015)
Neck carcinoma (esophageal squamous cell carcinoma)						↓																Sun et al. (2015)
Hepatocellular carcinomas	↓					↓						↓										Suren et al. (2014), Buhrmann et al. (2015), Kinugasa et al. (2007)
Metastatic melanoma	↑	↑																			↑	Bhat et al. (2012), Caruso et al. (2014), Resnick et al. (2005), Leinster et al. (2013)
Uterine			↑	↑																		Ebnet (2008), Choi et al. (2007)
Thyroid						↑																Aurrand-Lions et al. (2001)
Inflammatory bowel disease (IBD)	↑	↑																				Sun et al. (2011)
Neonatal ichthyosis-sclerosing cholangitis syndrome	↑																					He et al. (2013)
Icthyosis	↑																					He et al. (2013)
Experimental autoimmune encephalomyelitis (animal model of CNS inflammatory disease)					↓												↓					Dahiya et al. (2011)

Crohn's disease	↑↓	↑		↑↓	↓		↓				↑						↓	↓		↓		Pope et al. (2014), Bhat et al. (2015), Takehara et al. (2009), Huang et al. (2014)
Ulcerative colitis	↑	↑		↑↓											↑		↓	↓		↓		Sun et al. (2011), Takehara et al. (2009), Huang et al. (2014)
Pouchitis	↓	↑																↓				Hwang et al. (2014), Lin et al. (2013)
Chronic inflammatory pain (brain)			↑		↑												↓					Ikeo et al. (2015)
Multiple sclerosis (autoimmune encephalomyelitis)	↑				↑↓												↓	↓		↓		Satake et al. (2008)
Rheumatoid arthritis (RA)																	↓					Sobel et al. (2005)
Atherosclerosis	↑																↑	↑↓				Lee et al. (2005)
Amyotrophic lateral sclerosis					↓												↓	↓				Usami et al. (2006)
Acute lung inflammation		↓		↓	↓													↓				Swift et al. (1983)
Celiac disease		↑	↑	↑													↓	↓				Suh et al. (2013)
Collagenous colitis				↓													↓					Li et al. (2016))
Acute self-limited colitis (ASLC)				↑																		Zhang et al. (2013)

↑ Upregulated, ↓ downregulated, ↑↓ variable

Table 5.2 Studies of monoclonal antibodies directed against tight junction proteins

Condition	TJ target	Drug available (monoclonal antibodies)	References/clinical trial number
Colorectal cancer	Claudin-1	3A2, 6F6	Scharl et al. (2009), Zeissig et al. (2007)
Fibrosarcoma	Claudin-2	xi-1A2	Hadj-Rabia et al. (2004)
Various tumors, ovarian cancer	Claudin-3	KM3953, IgGH6	Denizot et al. (2012), Amasheh et al. (2009)
Bladder cancer, pancreatic and ovarian cancers	Claudin-4	KM3934, 4D3, 5D12, Clone 382321, KM3900	Weber et al. (2008), Landy et al. (2014), Brooks et al. (2005), Pfeiffer et al. (2011)
Breast and ovarian tumors, stomach cancer	Claudin-3, -4	KM3907, 5A5	Argaw et al. (2009), Prasad et al. (2005)
Ovarian cancer	Claudin-6	IMAB027	Nishioku et al. (2010)/ NCT03504397, NCT02054351
Gastric, gastro-esophageal cancer, and pancreatic cancer	Claudin-18.2	IMAB362 (claudiximab or zolbetuximab)	Chen et al. (2017), Garbuzova-Davis and Sanberg (2014),/ NCT01630083, NCT03504397, NCT03816163
Germ cell tumor	Claudin-6	IMAB027 (ASP1650)	NCT03760081
Prostate cancer	Claudin-11	DZ-50	Ohta et al. (2012)
Pancreatic or gastric including esophageal junction cancers	Claudin-18.2	CPO102	NCT05043987
Gastric or pancreatic cancer	Claudin-18.2	CT041 (CAR-T cells)	NCT04404595
Metastatic solid tumors	Claudin-18.2	TST001, SYSA1801, AB011, BNT141, MIL93, Q-1802, LM-102, CMG901, AMG 910	NCT04495296, NCT05009966, NCT04400383, NCT04683939, NCT04671875, NCT04856150, NCT05008445, NCT04805307, NCT04260191
IBD	Claudin-2	1A2	Khaleghi et al. (2016)

co-treatment with the anti-TNF-mAb (infliximab or adalimumab) exhibited an additive effect in restoring the TJ integrity (Khaleghi et al. 2016). Interestingly, in gastric cancer treatment, anti-claudin-18.2 mAb (claudiximab) has shown promising outcomes when added to standard chemotherapy by significantly increasing the median survival (Burgel et al. 2002).

Importantly, anti-CLDN mAbs are safe, and no relevant off-targets have been reported. Their marked therapeutic effects combined with excellent safety profiles are highly encouraging for their development in clinical applications.

5.4 Conclusion and Future Directions

In conclusion, the knowledge about the vast potential of TJs and their proteins in inflammation and cancer has evolved from the early findings. A myriad of studies has been highlighted demonstrating the dysregulation of tight junctional protein and its role in promoting specific cancers and attempted to address how altered expression of specific TJ proteins would actively contribute to tumor initiation, progression, and metastasis. The highlighted findings show the potential of targeting the TJ proteins, particularly claudins, for therapeutic application to act as novel therapeutic strategies for various cancers and IBD. Furthermore, with the promising preclinical and clinical trial data, the future of targeted cancer therapies shows an exciting path towards developing novel therapeutic molecules targeting TJs.

Declaration Conflict of Interest: The authors declare that there is no conflict of interest.

References

Alikanoglu AS et al (2015) Expression pattern and prognostic significance of claudin 1, 4 and 7 in pancreatic cancer. Asian Pac J Cancer Prev 16:4387–4392

Amasheh S et al (2009) Inflamed pouch mucosa possesses altered tight junctions indicating recurrence of inflammatory bowel disease. Int J Color Dis 24:1149–1156

Anderson JM, Van Itallie CM (2009) Physiology and function of the tight junction. Cold Spring Harb Perspect Biol 1:a002584

Arcangeli ML et al (2012) The junctional adhesion molecule-B regulates JAM-C-dependent melanoma cell metastasis. FEBS Lett 586:4046–4051

Argaw AT, Gurfein BT, Zhang Y, Zameer A, John GR (2009) VEGF-mediated disruption of endothelial CLN-5 promotes blood-brain barrier breakdown. Proc Natl Acad Sci U S A 106: 1977–1982

Aurrand-Lions M, Duncan L, Ballestrem C, Imhof BA (2001) JAM-2, a novel immunoglobulin superfamily molecule, expressed by endothelial and lymphatic cells. J Biol Chem 276:2733–2741

Bazzoni G (2011) Pathobiology of junctional adhesion molecules. Antioxid Redox Signal 15:1221–1234

Bazzoni G et al (2000) Interaction of junctional adhesion molecule with the tight junction components ZO-1, cingulin, and occludin. J Biol Chem 275:20520–20526

Bhat AA et al (2012) Caudal homeobox protein Cdx-2 cooperates with Wnt pathway to regulate claudin-1 expression in colon cancer cells. PLoS One 7:e37174

Bhat AA et al (2015) Claudin-7 expression induces mesenchymal to epithelial transformation (MET) to inhibit colon tumorigenesis. Oncogene 34:4570–4580

Bhat AA et al (2018) Tight junction proteins and signaling pathways in cancer and inflammation: a functional crosstalk. Front Physiol 9:1942

Bhat AA et al (2020) Claudin-1, A double-edged sword in cancer. Int J Mol Sci 21:569

Boireau S et al (2007) DNA-methylation-dependent alterations of claudin-4 expression in human bladder carcinoma. Carcinogenesis 28:246–258

Brennan K, Offiah G, McSherry EA, Hopkins AM (2010) Tight junctions: a barrier to the initiation and progression of breast cancer? J Biomed Biotechnol 2010:460607

Brooks TA, Hawkins BT, Huber JD, Egleton RD, Davis TP (2005) Chronic inflammatory pain leads to increased blood-brain barrier permeability and tight junction protein alterations. Am J Physiol Heart Circ Physiol 289:738–743

Bruewer M, Samarin S, Nusrat A (2006) Inflammatory bowel disease and the apical junctional complex. Ann N Y Acad Sci 1072:242–252

Buhrmann C et al (2015) Resveratrol induces chemosensitization to 5-fluorouracil through up-regulation of intercellular junctions, epithelial-to-mesenchymal transition and apoptosis in colorectal cancer. Biochem Pharmacol 98:51–68

Burgel N et al (2002) Mechanisms of diarrhea in collagenous colitis. Gastroenterology 123:433–443

Caruso M et al (2014) Claudin-1 expression is elevated in colorectal cancer precursor lesions harboring the BRAF V600E mutation. Transl Oncol 7:456–463

Chen L et al (2017) Monocytic cell junction proteins serve important roles in atherosclerosis via the endoglin pathway. Mol Med Rep 16:6750–6756

Choi YL et al (2007) Expression profile of tight junction protein claudin 3 and claudin 4 in ovarian serous adenocarcinoma with prognostic correlation. Histol Histopathol 22:1185–1195

Clayburgh DR, Shen L, Turner JR (2004) A porous defense: the leaky epithelial barrier in intestinal disease. Lab Investig 84:282–291

Cohen CJ et al (2001) The coxsackievirus and adenovirus receptor is a transmembrane component of the tight junction. Proc Natl Acad Sci U S A 98:15191–15196

Dahiya N, Becker KG, Wood WH, Zhang Y, Morin PJ (2011) Claudin-7 is frequently overexpressed in ovarian cancer and promotes invasion. PLoS One 6:e22119

Denizot J et al (2012) Adherent-invasive Escherichia coli induce claudin-2 expression and barrier defect in CEABAC10 mice and Crohn's disease patients. Inflamm Bowel Dis 18:294–304

Ebnet K (2008) Organization of multiprotein complexes at cell-cell junctions. Histochem Cell Biol 130:1–20

Ebnet K, Suzuki A, Ohno S, Vestweber D (2004) Junctional adhesion molecules (JAMs): more molecules with dual functions? J Cell Sci 117:19–29

Fanning AS, Jameson BJ, Jesaitis LA, Anderson JM (1998) The tight junction protein ZO-1 establishes a link between the transmembrane protein occludin and the actin cytoskeleton. J Biol Chem 273:29745–29753

Fedwick JP, Lapointe TK, Meddings JB, Sherman PM, Buret AG (2005) Helicobacter pylori activates myosin light-chain kinase to disrupt claudin-4 and claudin-5 and increase epithelial permeability. Infect Immun 73:7844–7852

Fink C et al (2006) Altered expression of ZO-1 and ZO-2 in Sertoli cells and loss of blood-testis barrier integrity in testicular carcinoma in situ. Neoplasia 8:1019–1027

Forster C (2008) Tight junctions and the modulation of barrier function in disease. Histochem Cell Biol 130:55–70

French AD et al (2009) PKC and PKA phosphorylation affect the subcellular localization of claudin-1 in melanoma cells. Int J Med Sci 6:93–101

Furuse M et al (1993) Occludin: a novel integral membrane protein localizing at tight junctions. J Cell Biol 123:1777–1788

Furuse M, Sasaki H, Fujimoto K, Tsukita S (1998a) A single gene product, claudin-1 or -2, reconstitutes tight junction strands and recruits occludin in fibroblasts. J Cell Biol 143:391–401

Furuse M, Fujita K, Hiiragi T, Fujimoto K, Tsukita S (1998b) Claudin-1 and -2: novel integral membrane proteins localizing at tight junctions with no sequence similarity to occludin. J Cell Biol 141:1539–1550

Garbuzova-Davis S, Sanberg PR (2014) Blood-CNS barrier impairment in ALS patients versus an animal model. Front Cell Neurosci 8:21

Garrod D, Chidgey M (2008) Desmosome structure, composition and function. Biochim Biophys Acta 1778:572–587
Gasbarrini G, Montalto M (1999) Structure and function of tight junctions. Role in intestinal barrier. Ital J Gastroenterol Hepatol 31:481–488
Ghislin S et al (2011) Junctional adhesion molecules are required for melanoma cell lines transendothelial migration in vitro. Pigment Cell Melanoma Res 24:504–511
Gonzalez-Mariscal L, Betanzos A, Nava P, Jaramillo BE (2003) Tight junction proteins. Prog Biophys Mol Biol 81:1–44
Goodenough DA, Paul DL (2009) Gap junctions. Cold Spring Harb Perspect Biol 1:a002576
Grone J et al (2007) Differential expression of genes encoding tight junction proteins in colorectal cancer: frequent dysregulation of claudin-1, -8 and -12. Int J Color Dis 22:651–659
Hadj-Rabia S et al (2004) Claudin-1 gene mutations in neonatal sclerosing cholangitis associated with ichthyosis: a tight junction disease. Gastroenterology 127:1386–1390
Hartsock A, Nelson WJ (2008) Adherens and tight junctions: structure, function and connections to the actin cytoskeleton. Biochim Biophys Acta 1778:660–669
Haynes MD et al (2005) Tight junctions and bladder cancer (review). Int J Mol Med 16:3–9
He ZY et al (2013) Folate-linked lipoplexes for short hairpin RNA targeting claudin-3 delivery in ovarian cancer xenografts. J Control Release 172:679–689
Hewitt KJ, Agarwal R, Morin PJ (2006) The claudin gene family: expression in normal and neoplastic tissues. BMC Cancer 6:186
Hollander D (1988) Crohn's disease–a permeability disorder of the tight junction? Gut 29:1621–1624
Hollander D (1999) Clinician's guide through the tight junctions. Ital J Gastroenterol Hepatol 31:435–439
Hollander D et al (1986) Increased intestinal permeability in patients with Crohn's disease and their relatives. A possible etiologic factor. Ann Intern Med 105:883–885
Hoover KB, Liao SY, Bryant PJ (1998) Loss of the tight junction MAGUK ZO-1 in breast cancer: relationship to glandular differentiation and loss of heterozygosity. Am J Pathol 153:1767–1773
Huang J et al (2014) The expression of claudin 1 correlates with beta-catenin and is a prognostic factor of poor outcome in gastric cancer. Int J Oncol 44:1293–1301
Hwang TL, Changchien TT, Wang CC, Wu CM (2014) Claudin-4 expression in gastric cancer cells enhances the invasion and is associated with the increased level of matrix metalloproteinase-2 and -9 expression. Oncol Lett 8:1367–1371
Ikenouchi J et al (2005) Tricellulin constitutes a novel barrier at tricellular contacts of epithelial cells. J Cell Biol 171:939–945
Ikeo K et al (2015) Junctional adhesion molecule-A promotes proliferation and inhibits apoptosis of gastric cancer. Hepato-Gastroenterology 62:540–545
Ip YC, Cheung ST, Lee YT, Ho JC, Fan ST (2007) Inhibition of hepatocellular carcinoma invasion by suppression of claudin-10 in HLE cells. Mol Cancer Ther 6:2858–2867
Itoh M, Bissell MJ (2003) The organization of tight junctions in epithelia: implications for mammary gland biology and breast tumorigenesis. J Mammary Gland Biol Neoplasia 8:449–462
Kaihara T et al (2003) Dedifferentiation and decreased expression of adhesion molecules, E-cadherin and ZO-1, in colorectal cancer are closely related to liver metastasis. J Exp Clin Cancer Res 22:117–123
Khaleghi S, Ju JM, Lamba A, Murray JA (2016) The potential utility of tight junction regulation in celiac disease: focus on larazotide acetate. Ther Adv Gastroenterol 9:37–49
Kim TH et al (2008) Down-regulation of claudin-2 in breast carcinomas is associated with advanced disease. Histopathology 53:48–55
Kimura Y et al (1997) Expression of occludin, tight-junction-associated protein, in human digestive tract. Am J Pathol 151:45–54
Kinugasa T et al (2007) Selective up-regulation of claudin-1 and claudin-2 in colorectal cancer. Anticancer Res 27:3729–3734

Kohno Y et al (2006) Expression of claudin7 is tightly associated with epithelial structures in synovial sarcomas and regulated by an Ets family transcription factor, ELF3. J Biol Chem 281: 38941–38950
Kominsky SL et al (2003) Loss of the tight junction protein claudin-7 correlates with histological grade in both ductal carcinoma in situ and invasive ductal carcinoma of the breast. Oncogene 22: 2021–2033
Kowalczyk AP, Green KJ (2013) Structure, function, and regulation of desmosomes. Prog Mol Biol Transl Sci 116:95–118
Landy J et al (2014) Innate immune factors in the development and maintenance of pouchitis. Inflamm Bowel Dis 20:1942–1949
Landy J et al (2016) Tight junctions in inflammatory bowel diseases and inflammatory bowel disease associated colorectal cancer. World J Gastroenterol 22:3117–3126
Langer HF et al (2011) A novel function of junctional adhesion molecule-C in mediating melanoma cell metastasis. Cancer Res 71:4096–4105
Lauko A, Mu Z, Gutmann DH, Naik UP, Lathia JD (2020) Junctional adhesion molecules in cancer: a paradigm for the diverse functions of cell-cell interactions in tumor progression. Cancer Res 80:4878–4885
Lee JW et al (2005) Increased expressions of claudin-1 and claudin-7 during the progression of cervical neoplasia. Gynecol Oncol 97:53–59
Leinster DA et al (2013) Endothelial cell junctional adhesion molecule C plays a key role in the development of tumors in a murine model of ovarian cancer. FASEB J 27:4244–4253
Li CP et al (2016) CLDN14 is epigenetically silenced by EZH2-mediated H3K27ME3 and is a novel prognostic biomarker in hepatocellular carcinoma. Carcinogenesis 37:557–566
Lin Z et al (2013) The distinct expression patterns of claudin-2, -6, and -11 between human gastric neoplasms and adjacent non-neoplastic tissues. Diagn Pathol 8:133
Liu Y et al (2000) Human junction adhesion molecule regulates tight junction resealing in epithelia. J Cell Sci 113(Pt 13):2363–2374
Ma X et al (2015) Claudin-4 controls the proliferation, apoptosis, migration and in vivo growth of MCF-7 breast cancer cells. Oncol Rep 34:681–690
Majer A, Blanchard AA, Medina S, Booth SA, Myal Y (2016) Claudin 1 expression levels affect miRNA dynamics in human basal-like breast cancer cells. DNA Cell Biol 35:328–339
Martin TA, Mansel RE, Jiang WG (2002) Antagonistic effect of NK4 on HGF/SF induced changes in the transendothelial resistance (TER) and paracellular permeability of human vascular endothelial cells. J Cell Physiol 192:268–275
Martin TA, Mansel RE, Jiang WG (2010) Loss of occludin leads to the progression of human breast cancer. Int J Mol Med 26:723–734
Martin-Padura I et al (1998) Junctional adhesion molecule, a novel member of the immunoglobulin superfamily that distributes at intercellular junctions and modulates monocyte transmigration. J Cell Biol 142:117–127
Matter K, Aijaz S, Tsapara A, Balda MS (2005) Mammalian tight junctions in the regulation of epithelial differentiation and proliferation. Curr Opin Cell Biol 17:453–458
Mese G, Richard G, White TW (2007) Gap junctions: basic structure and function. J Invest Dermatol 127:2516–2524
Morin PJ (2005) Claudin proteins in human cancer: promising new targets for diagnosis and therapy. Cancer Res 65:9603–9606
Morohashi S et al (2007) Decreased expression of claudin-1 correlates with recurrence status in breast cancer. Int J Mol Med 20:139–143
Munkholm P et al (1994) Intestinal permeability in patients with Crohn's disease and ulcerative colitis and their first degree relatives. Gut 35:68–72
Nelson WJ (2003) Adaptation of core mechanisms to generate cell polarity. Nature 422:766–774
Nishioku T et al (2010) Disruption of the blood-brain barrier in collagen-induced arthritic mice. Neurosci Lett 482:208–211

Ohta H, Chiba S, Ebina M, Furuse M, Nukiwa T (2012) Altered expression of tight junction molecules in alveolar septa in lung injury and fibrosis. Am J Physiol Lung Cell Mol Physiol 302: 193–205

Orban E et al (2008) Different expression of occludin and ZO-1 in primary and metastatic liver tumors. Pathol Oncol Res 14:299–306

Osanai M et al (2006) Epigenetic silencing of occludin promotes tumorigenic and metastatic properties of cancer cells via modulations of unique sets of apoptosis-associated genes. Cancer Res 66:9125–9133

Osanai M, Takasawa A, Murata M, Sawada N (2017) Claudins in cancer: bench to bedside. Pflugers Arch 469:55–67

Paschoud S, Bongiovanni M, Pache JC, Citi S (2007) Claudin-1 and claudin-5 expression patterns differentiate lung squamous cell carcinomas from adenocarcinomas. Mod Pathol 20:947–954

Pfeiffer F et al (2011) Claudin-1 induced sealing of blood-brain barrier tight junctions ameliorates chronic experimental autoimmune encephalomyelitis. Acta Neuropathol 122:601–614

Polette M et al (2007) Beta-catenin and ZO-1: shuttle molecules involved in tumor invasion-associated epithelial-mesenchymal transition processes. Cells Tissues Organs 185:61–65

Pope JL et al (2014) Claudin-1 overexpression in intestinal epithelial cells enhances susceptibility to adenamatous polyposis coli-mediated colon tumorigenesis. Mol Cancer 13:167

Prasad S et al (2005) Inflammatory processes have differential effects on claudins 2, 3 and 4 in colonic epithelial cells. Lab Investig 85:1139–1162

Rachow S et al (2013) Occludin is involved in adhesion, apoptosis, differentiation and Ca2+-homeostasis of human keratinocytes: implications for tumorigenesis. PLoS One 8:e55116

Ram AK, Pottakat B, Vairappan B (2018) Increased systemic zonula occludens 1 associated with inflammation and independent biomarker in patients with hepatocellular carcinoma. BMC Cancer 18:572

Rangel LB et al (2003) Tight junction proteins claudin-3 and claudin-4 are frequently overexpressed in ovarian cancer but not in ovarian cystadenomas. Clin Cancer Res 9:2567–2575

Raskov H, Burcharth J, Pommergaard HC (2017) Linking gut microbiota to colorectal cancer. J Cancer 8:3378–3395

Resnick MB, Konkin T, Routhier J, Sabo E, Pricolo VE (2005) Claudin-1 is a strong prognostic indicator in stage II colonic cancer: a tissue microarray study. Mod Pathol 18:511–518

Saitou M et al (2000) Complex phenotype of mice lacking occludin, a component of tight junction strands. Mol Biol Cell 11:4131–4142

Salvador E, Burek M, Forster CY (2016) Tight junctions and the tumor microenvironment. Curr Pathobiol Rep 4:135–145

Sanchez-Pulido L, Martin-Belmonte F, Valencia A, Alonso MA (2002) MARVEL: a conserved domain involved in membrane apposition events. Trends Biochem Sci 27:599–601

Santin AD et al (2007) Overexpression of claudin-3 and claudin-4 receptors in uterine serous papillary carcinoma: novel targets for a type-specific therapy using Clostridium perfringens enterotoxin (CPE). Cancer 109:1312–1322

Satake S et al (2008) Cdx2 transcription factor regulates claudin-3 and claudin-4 expression during intestinal differentiation of gastric carcinoma. Pathol Int 58:156–163

Sauer T, Pedersen MK, Ebeltoft K, Naess O (2005) Reduced expression of Claudin-7 in fine needle aspirates from breast carcinomas correlate with grading and metastatic disease. Cytopathology 16:193–198

Sawada N (2013) Tight junction-related human diseases. Pathol Int 63:1–12

Scharl M, Paul G, Barrett KE, McCole DF (2009) AMP-activated protein kinase mediates the interferon-gamma-induced decrease in intestinal epithelial barrier function. J Biol Chem 284: 27952–27963

Schmidt A, Utepbergenov DI, Krause G, Blasig IE (2001) Use of surface plasmon resonance for real-time analysis of the interaction of ZO-1 and occludin. Biochem Biophys Res Commun 288: 1194–1199

Schneeberger EE, Lynch RD (2004) The tight junction: a multifunctional complex. Am J Physiol Cell Physiol 286:1213–1228

Schulzke JD et al (2005) Epithelial transport and barrier function in occludin-deficient mice. Biochim Biophys Acta 1669:34–42

Sharma RK et al (2016) A spontaneous metastasis model reveals the significance of claudin-9 overexpression in lung cancer metastasis. Clin Exp Metastasis 33:263–275

Shin K, Fogg VC, Margolis B (2006) Tight junctions and cell polarity. Annu Rev Cell Dev Biol 22: 207–235

Singh AB, Dhawan P (2015) Claudins and cancer: fall of the soldiers entrusted to protect the gate and keep the barrier intact. Semin Cell Dev Biol 42:58–65

Sobel G et al (2005) Increased expression of claudins in cervical squamous intraepithelial neoplasia and invasive carcinoma. Hum Pathol 36:162–169

Soini Y (2005) Expression of claudins 1, 2, 3, 4, 5 and 7 in various types of tumours. Histopathology 46:551–560

Suh Y et al (2013) Claudin-1 induces epithelial-mesenchymal transition through activation of the c-Abl-ERK signaling pathway in human liver cells. Oncogene 32:4873–4882

Sun C et al (2011) Efficient inhibition of ovarian cancer by short hairpin RNA targeting claudin-3. Oncol Rep 26:193–200

Sun L et al (2015) A novel role of OS-9 in the maintenance of intestinal barrier function from hypoxia-induced injury via p38-dependent pathway. Int J Biol Sci 11:664–671

Suren D et al (2014) Loss of tight junction proteins (Claudin 1, 4, and 7) correlates with aggressive behavior in colorectal carcinoma. Med Sci Monit 20:1255–1262

Swift JG, Mukherjee TM, Rowland R (1983) Intercellular junctions in hepatocellular carcinoma. J Submicrosc Cytol 15:799–810

Tabaries S, Siegel PM (2017) The role of claudins in cancer metastasis. Oncogene 36:1176–1190

Takehara M, Nishimura T, Mima S, Hoshino T, Mizushima T (2009) Effect of claudin expression on paracellular permeability, migration and invasion of colonic cancer cells. Biol Pharm Bull 32:825–831

Teshima CW, Dieleman LA, Meddings JB (2012) Abnormal intestinal permeability in Crohn's disease pathogenesis. Ann N Y Acad Sci 1258:159–165

Tobioka H et al (2004) Occludin expression decreases with the progression of human endometrial carcinoma. Hum Pathol 35:159–164

Tsukita S, Furuse M (1999) Occludin and claudins in tight-junction strands: leading or supporting players? Trends Cell Biol 9:268–273

Tsukita S, Furuse M (2000) Pores in the wall: claudins constitute tight junction strands containing aqueous pores. J Cell Biol 149:13–16

Tsukita S, Tsukita S (1989) Isolation of cell-to-cell adherens junctions from rat liver. J Cell Biol 108:31–41

Tsukita S, Yamazaki Y, Katsuno T, Tamura A, Tsukita S (2008) Tight junction-based epithelial microenvironment and cell proliferation. Oncogene 27:6930–6938

Tzelepi VN, Tsamandas AC, Vlotinou HD, Vagianos CE, Scopa CD (2008) Tight junctions in thyroid carcinogenesis: diverse expression of claudin-1, claudin-4, claudin-7 and occludin in thyroid neoplasms. Mod Pathol 21:22–30

Usami Y et al (2006) Reduced expression of claudin-7 correlates with invasion and metastasis in squamous cell carcinoma of the esophagus. Hum Pathol 37:569–577

Vetrano S, Danese S (2009) The role of JAM-A in inflammatory bowel disease: unrevealing the ties that bind. Ann N Y Acad Sci 1165:308–313

Weber CR, Turner JR (2007) Inflammatory bowel disease: is it really just another break in the wall? Gut 56:6–8

Weber CR, Nalle SC, Tretiakova M, Rubin DT, Turner JR (2008) Claudin-1 and claudin-2-expression is elevated in inflammatory bowel disease and may contribute to early neoplastic transformation. Lab Investig 88:1110–1120

Zeissig S et al (2007) Changes in expression and distribution of claudin 2, 5 and 8 lead to discontinuous tight junctions and barrier dysfunction in active Crohn's disease. Gut 56:61–72

Zhang Z et al (2013) Expression of CLDN1 and CLDN10 in lung adenocarcinoma in situ and invasive lepidic predominant adenocarcinoma. J Cardiothorac Surg 8:95

Zhang X et al (2019) Decreased expression of ZO-1 is associated with tumor metastases in liver cancer. Oncol Lett 17:1859–1864

Zhu L et al (2019) Claudin family participates in the pathogenesis of inflammatory bowel diseases and colitis-associated colorectal cancer. Front Immunol 10:1441

Tight Junctions and Signaling Pathways in Cancer

6

Sana Khurshid, Burhan UlHaq, Sadaf Khursheed, Hana Q. Sadida, Tariq Masoodi, Mayank Singh, Ammira S. Al-Shabeeb Akil, Ajaz A. Bhat, and Muzafar A. Macha

Abstract

Tight junctions (TJs) are intercellular connections that close the gap between the individual cells and limit the paracellular entry of harmful antigens, toxins, microbes, etc. In addition, they also regulate vesicle trafficking, signal transduction, transcription, and cytoskeletal dynamics. In contrast to other specialized cell connections such as adherens, gap junctions, and desmosomes, TJs form an uninterrupted intercellular contact at the apical-most end of the lateral side of epithelial cells. Compared to other junctions TJs are made up of different proteins, including claudins, which form the distinctive network of interconnected strands; the multi-PDZ proteins ZO-1, ZO-2, ZO-3, MAGI-1,

S. Khurshid · M. A. Macha (✉)
Watson-Crick Centre for Molecular Medicine, Islamic University of Science and Technology, Awantipora, Jammu and Kashmir, India

B. UlHaq
Department of Zoology, School of Life Sciences, Central University of Kashmir, Ganderbal, Jammu and Kashmir, India

S. Khursheed
Department of Microbiology, Sher-I-Kashmir Institute of Medical Science (SKIMS), Soura, Jammu and Kashmir, India

H. Q. Sadida · A. S. A.-S. Akil · A. A. Bhat
Department of Human Genetics-Precision Medicine in Diabetes, Obesity and Cancer Research Program, Sidra Medicine, Doha, Qatar

T. Masoodi
Laboratory of Cancer Immunology and Genetics, Sidra Medicine, Doha, Qatar

M. Singh
Department of Medical Oncology, Dr. BRAIRCH, All India Institute of Medical Sciences, New Delhi, India

A. A. Bhat et al. (eds.), *Tight Junctions in Inflammation and Cancer*,
https://doi.org/10.1007/978-981-99-2415-8_6

PatJ, MUPP1, and PALS1, which are interconnected with the transmembrane claudin proteins and are hypothesized to cluster the barrier-forming proteins and enable redundant attachments to the actin cytoskeleton. Disruption of TJ barriers results in the deregulated passage of materials and toxins through the epithelia contributing to a vast array of pathologies. In addition, TJs of endothelial and epithelial cells act as a hub for signaling pathways that determine the proliferative and metastatic ability in cancers. Claudins' function in cancer signaling has been significantly researched. Numerous proof-of-principle studies targeting certain claudins for chemotherapy have been conducted due to the massive overexpression of particular claudins in distinct cancer types.

Keywords

Claudins · Cancer · Cell adhesion · Cell proliferation · Inflammation · Signaling network · Tight junctions

6.1 Introduction

Multicellular organisms primarily require developing a unique internal environment to sustain their existence. Epithelial and endothelial cells are essential in lining the body surfaces. Epithelial cells cover all the surfaces of the skin, GI tract, respiratory tract, etc., or line a cavity, and endothelial cells line vasculature to perform diverse functions. For example, endothelial and epithelial sheets work efficiently as a barrier, protect organs from their surroundings, and aid in sustaining homeostasis (Tsukita et al. 2001; Tsukita and Furuse 2002). This structural stability and barrier function of endothelium and epithelium is maintained through the cooperation of various proteins comprising tight junctions (TJs), anchoring junctions or adheren junctions (AJs), desmosomes, and gap junctions (GJs) (Kottke et al. 2006). These proteins primarily regulate intercellular communication and paracellular transport among the cells in the tissues (Bhat et al. 1942) (Fig. 6.1). Proper adhesion of cell surface receptors is necessary for the optimal function of these proteins.

TJs regulate the barrier function by inhibiting the free diffusion of solutes through paracellular spaces (Bhat et al. 1942). The primary building blocks of TJs are occludins, claudins, and junctional adhesion molecules (JAM), which will be mentioned in more detail in this book chapter (González-Mariscal et al. 2003). AJ proteins mainly involved in cell surface adhesion include cadherins, catenins, and integrins (Chattopadhyay et al. 2003).

AJs are often formed between cells and play a crucial role in the hemostasis of tissues and development (Kottke et al. 2006). Cytoskeleton-bound AJs provide support and act as signaling hubs important for gene regulation (Meng and Takeichi 2009). While cadherins on the membranes of nearby cells aid in connecting cells to each other (Dejana 2004), catenins like β-catenin help cell adhesion and cell growth and differentiation and, more importantly, play an essential part in the transformation of normal cells (Zhan et al. 2017). Gap junctions (GJs), a family of transmembrane proteins, aid cell–cell communication by transporting and directly exchanging

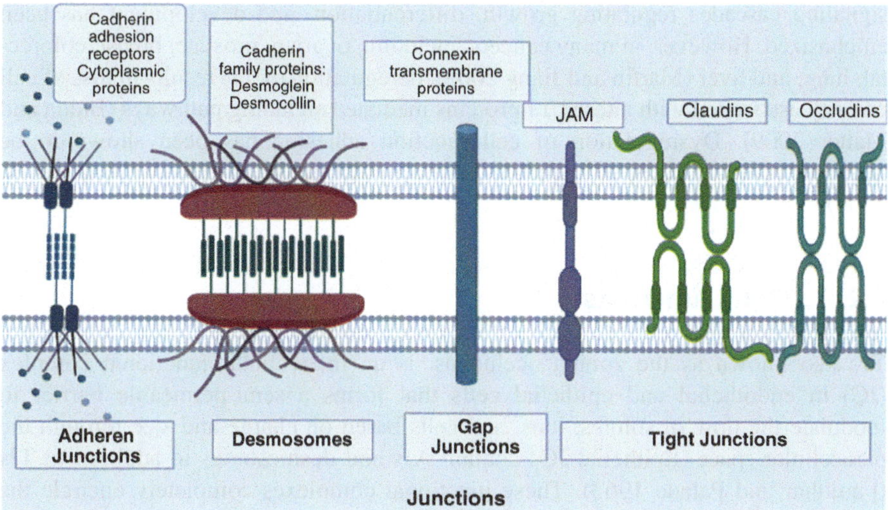

Fig. 6.1 Epithelial intracellular junction. Schematic representation of the epithelial junction in the vertebrate cell. TJs, adherens junctions, and gap junctions are located in the cell's apical region, whereas desmosomes are located toward the basal area

molecules and solutes between cells through a set of integral membrane proteins called connexins (Dbouk et al. 2009). It has been shown that GJs serve critical roles in tissue homeostasis, growth, and development when functioning normally or adequately (Kandouz and Batist 2010).

TJs have long been the most mysterious of all adhesion complexes (Zihni et al. 2016). TJs act as barriers for the cell with restricted entry, maintaining cell polarity. Active transport is the only possible transport in a standard, functioning, healthy cell through a TJ (Anderson and Van Itallie 2009). Since TJ dysregulation results in altered barrier function, affecting the levels of inflammatory cytokines such as IFN-α, IFN-γ, IL-6, and IL-1β, as found in associated inflammatory illnesses such as inflammatory bowel disease (IBD), multiple sclerosis, and cancer (Bhat et al. 1942). Deregulation of TJs in the gastrointestinal tract (GI) and biliary tract epithelial cells results in many diseases, including diarrhea and jaundice. In addition, dysfunction of TJs results in edema formation due to altered vascular permeability of caveolar transcellular and paracellular pathways of endothelial sheets.

Additionally, the TJ proteins are essential for maintaining the intestinal epithelium's integrity. Any alteration like gut inflammation causes the intestinal epithelium to be disrupted, as is the case with IBD, such as Crohn's disease or ulcerative colitis (UC) (Bhat et al. 1942). Epithelial cells have basolateral and apical membranes with different lipid and protein compositions and differ in function. TJs among these surfaces surround the apical cellular poles and prevent the intermixing of molecules among basolateral and apical membranes. However, disruption of this function fails vectorial work and causes loss of cell polarity. With increasing functional and structural knowledge of TJ biology, their importance in cellular signaling cascades regulating growth, differentiation, and development has been emphasized. However, in many cancers, including ovarian, prostate, breast, colorectal, lung, and liver (Martin and Jiang 2009), deregulation of TJs results in loss of cell polarity associated with altered TJ proteins mediated signaling pathways (Balda and Matter 2009). Dysregulation of cell junction adhesion has been shown to be intensely involved in the epithelial-to-mesenchymal transition (EMT) (Morris et al. 2008).

6.2 Tight Junctions

TJ, also known as the zonula occludens, is an intercellular junctional complex (JC) in endothelial and epithelial cells that forms a semi-permeable barrier to modulate the flow of solutes, ions, and cells based on charge and size through the paracellular space. Epithelial JCs contain AJs and desmosomes in addition to TJs (Farquhar and Palade 1963). These junctional complexes completely encircle the cells, resulting in a continuous junctional belt that interconnects with cells in close vicinity. TJs are the most apical part of the junctional complex in epithelial cells, whereas, in endothelia, their localization is more diverse (Cereijido et al. 2000; Dejana et al. 2009). TJs have been identified by freeze-fracture electron microscopy as a collection of anastomosing membrane strands that live in close contact sites between the outer leaflets of the neighboring plasma membranes, effectively occluding the intercellular space between adjacent cells (Tsukita et al. 2001; Cereijido et al. 1998). Thus, based upon these morphological observations and previous research, TJs have been typically demonstrated to either create a semi-permeable barrier that is capable of finely modulating the movement of ions, solutes, and cells via the paracellular space and that differ based on size and charge or as a fence that takes part in the maintenance and establishment of apicobasal polarity.

6.2.1 Molecular Composition of Tight Junctions

The molecular structure of TJs has been extensively studied. TJ complexes are made up of integral membrane proteins that are connected to an adaptor protein network, forming a cytoplasmic plaque. Claudins are a class of essential membrane proteins that contain TJ strands and regulate paracellular permeability. Claudins have a cytoplasmic region that interacts with the zonula occludens (ZO) family of scaffolding proteins, which are required for TJ assembly. In addition to claudins, membrane proteins such as TJ-associated MARVEL domain-containing proteins (tricellulin, MarvelD3, TAMPs: occludin) and a single-span transmembrane proteins, involving the junctional adhesion molecule (JAM) from the immunoglobulin super family localize to TJs and play significant roles in TJ formation and function.

Claudins and occludins are the two most significant TJ proteins that regulate the crucial component of the cell. Along with other TJ proteins, claudins are a family of transmembrane proteins essential for epithelial TJs to function correctly. The claudins can hold all the proteins of the cytoplasmic milieu, along with their exquisite interaction with occludins (Bhat et al. 1942). The claudins are significant because they are primarily responsible for the formation of TJs (Cording et al. 2013). Members of the closed group are claudins 1, 3, 5, 11, 14, and 19 and are responsible for maintaining the cell's water-tight integrity. Claudins 10a and 17 allow anions, while claudins 10 b and 15 permit cations to pass through. Claudin-2 is permeable to anions and water, providing proper water availability and ions for cellular functions (Günzel and Fromm 2012). Cingulin, Pals1 (proteins associated with Lin Seven 1), MUPP1 (multi-PDZ domain protein 1), as well as ZO1, ZO-2, ZO-3 (Zona occludens) are the additional TJ proteins that link transmembrane proteins to the actin cytoskeleton. Src homology 3 (SH3) domains, N-terminal region with 3PDZ fields, and guanylate kinase (GUK) domain are the three ZO-1 proteins with shared structural characteristics. ZO proteins form the primary network for protein interactions. ZO-1/2-deficient cells do not form TJs because ZO-1 and ZO-2 are essential for TJ formation (Ando-Akatsuka et al. 1999). ZO proteins are multi-domain scaffolding proteins that form oligomers and interact with the actin cyto-skeleton, TJ-associated membrane proteins, signaling proteins, and AJ proteins, including catenin and afadin (Itoh et al. 1997, 1999, 2012; Furuse et al. 1994; Ebnet et al. 2000; Bazzoni et al. 2000). The C-termini of claudins directly binds with all ZO proteins of the first PDZ domain, which is considered important for TJ formation and function (Bhat et al. 1942; Otani and Furuse 2020). Multiple TJ constituents, involving claudin and occludin, can be recruited by the ZO family of proteins, creating phase-separated droplets (Beutel et al. 2019; Schwayer et al. 2019). Therefore phase separation of these proteins can be caused by the recruitment of the ZO protein family to primordial AJs and allowing TJ-incorporated membrane proteins to be recruited to generate TJ assembly (Beutel et al. 2019).

6.2.2 Structure of Tight Junctions

A stable multiprotein complex composed of peripheral and integral membrane proteins is the most common model of TJ structure. Transmembrane proteins, such as claudins, occludin, tricellulin, JAM, Coxsackie, and adenovirus-associated receptors, are the principal categories of integral membrane proteins. The discovery of claudins as critical components of TJ strands was a defining moment in our knowledge of TJ structure. The unique shape of TJ strands is assumed to be based on lateral linkage between claudin molecules inside the plasma membrane, along with homotypic adhesive contacts between claudin molecules on neighboring cells. Because claudins allow selective ion diffusion across TJs, a hypothetical model of tight junction organization appeared in which several claudin-based pores allow the passage of ions through a membrane contact site that is otherwise sealed. However, the actual molecular structure of such paracellular pores is still unknown.

The TJ structure becomes more complex by binding other essential membrane proteins to the claudin-based strands. The first transmembrane component of TJs idenitified was occludin, which likewise localizes to the TJ strands and has been

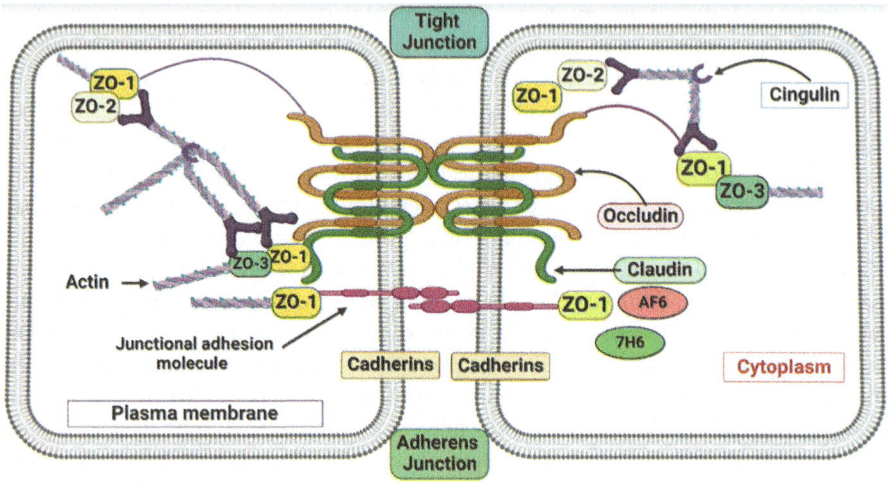

Fig. 6.2 Systemic illustration of the basic structure of transmembrane components of tight junction; ZO-1 or ZO-2 is required for the assembly of claudins and occludin, which results in the formation of tight junctional strands. The role of the other scaffolding proteins ZO-3/MAGI/MUP1 is unclear. ZOs and cingulin can directly link the actin cytoskeleton

connected to the regulation of size-selective diffusion in TJs. JAMs and related proteins function as homotypical and heterotypic adhesion proteins, regulating processes such as leukocyte transmigration. The cytoplasmic plaque, which lies beneath the membrane domain, is a densely packed meshwork of peripheral proteins that connects the integral membrane proteins to the underlying actin cytoskeleton and several signaling proteins. Notable examples include the zona occludens proteins ZO-1, -2, and -3, which possess several protein-interaction domains, including three PDZ domains and an SH3 domain, and due to which they exhibit an affinity for the number of cytoskeletal proteins, membrane proteins, and signaling molecules. Numerous cytosolic adaptor proteins can interact with other TJ components and form various protein complexes. Moreover, TJ components can bind with proteins of adherens junctions (e.g. afadin/AF- 6), ZO-1 with α-catenin, a kind of complex formation that can be seen in cells that possess only adherens junctions or more transiently, during junction assembly. This complicated network of protein interactions is considered important for properly organizing the integral membrane components of the junction, transfer of signals to the cell interior, and modulation of junction assembly and function (Steed et al. 2010) (Fig. 6.2).

6.2.3 Regulation of Tight Junctions

Although the integrity of the actin cytoskeleton and ATP is needed for the maintenance and development of TJ, the barrier function depends more on actin and ATP than the fence function (Sawada et al. 2003). The modulation of RhoAGTPase activity by Na, K-ATPase is essential for forming TJ and other intercellular connections in polarized epithelial cells. TGF-β and IL-10 protect the barrier function against cytokine-induced deterioration, even though several factors cause the barrier function to deteriorate. At the transcriptional level, tumor necrosis factor-α (TNF-α) and interferon-α (IFN-α) suppress occludin expression (Mankertz et al. 2000). Other variables, including the mitogen-activated protein kinase (MAPK) pathway, may affect the operation of tight-junction barriers through diverse signal transduction systems. Hepatocyte nuclear factor-4 (HNF-4) is a transcriptional factor that has been demonstrated to initiate the expression of at least three TJ molecules in F9 cells, including occludin, claudin-6, and claudin-7, also the biogenesis of functional TJs and the acquisition of polarized epithelial morphology (Ooshio et al. 2010; Kojima et al. 2007).

6.3 Function of Tight Junctions

To maintain cell polarity, TJs, the apical-most part of intercellular junctional complexes, isolate the apical from basolateral cell-surface domains (the fence function) while simultaneously regulating solute and water transport across the paracellular space (the barrier function). The TJs fence and barrier functions are both compartmentalized, with the fence function happening at the subcellular level and the barrier function occurring at the organ level, respectively.

6.3.1 Gate Function of Tight Junctions

Epithelia serve as barriers to compartmentalizing the body. However, the permeability of epithelia differs depending on their function. The two essential functions of TJs are the determination of selective permeability in leaky epithelia and the development of a permeability barrier in tight epithelia (Zihni et al. 2016; Anderson and Van Itallie 2009; Shen et al. 2011). At least two mechanisms govern paracellular permeability: a charge-selective small-pore pathway with an estimated diameter of 4 microns and a size-selective pathway that allows molecules up to 60 microns to pass through (Zihni et al. 2016; Anderson and Van Itallie 2009; Shen et al. 2011; Watson et al. 2005). The discovery of claudins leads the way to understanding the molecular basis of TJ permselectivity. In 1999, claudin-16/paracellin was discovered to be the gene responsible for familial renal hypomagnesemia, demonstrating that claudins modulate ion paracellular conductance (Simon et al. 1999). One more study based on this concept showed that overexpression of claudin-2 significantly increased the ion conductance of epithelia, converting a tight epithelium into a leaky epithelium (Furuse et al. 2001). A recent study has demonstrated that claudin-2 forms cation-selective paracellular channels (Amasheh et al. 2002; Colegio et al. 2002). The permselectivity is determined by charged residues in the first extracellular domain of claudins, which is clarified by claudin crystal structure and site-directed mutagenesis (Colegio et al. 2002; Suzuki et al. 2014, 2015; Yu et al. 2009). A recent patch-clamp research showed that paracellular channels that depend on claudin-2 are dynamically gated (Weber et al. 2015). Thus these findings illustrate that claudins are responsible for forming the charge-selective small-pore pathway that modulates ion conductance. In claudin quintuple-KO MDCK II cells, the permeability barrier against small molecules and ions up to 4KDa is disrupted. Still, it maintains the molecular permeability, revealing that the macromolecule permeability barrier can form without claudin-based TJ strands (Otani et al. 2019).

JAMs have been implicated in epithelial barrier function. Claudin-5 KO endothelial cells and claudin quintuple-KO MDCK II cells retain close membrane apposition. The intercellular space expanded, and the macromolecule permeability barrier for molecules larger than 4kD was disrupted when JAM-A was further deleted from claudin quintuple-KO MDCK II cells, indicating the significance of JAMs in the development of the macromolecule permeability barrier (Otani et al. 2019). Considering the intriguing similarity between the intermembrane distance in claudin quintuple-KO MDCK II cells (6–7 nm) and the pore size of the size-selective large-pore pathway (~60 Å), it is tempting to speculate that close membrane apposition acts as a molecular sieve that physically inhibits the passage of larger macromolecules. Consistent with this idea, there are some examples of macromolecule permeability barriers forming without the TJ strand. Adherens junctions (AJs) have been hypothesized to have the ability to act as a permeability barrier for macromolecules (Fujisawa et al. 1976; Smith and Reese 2016; Bunt-Milam et al. 1985). Because AJs and JAM-mediated contacts have different intermembrane distances, the adhesion molecules may be able to tune the size dependency of the macromolecule permeability barrier. In addition to JAMs and AJs, macromolecule permeability barrier formation has been implicated in occludin and tricellulin, despite the role of occludin being under debate (Balda et al. 1996; Al-Sadi et al. 2011; Richter et al. 2019; Krug et al. 2009). The leak route occurs when actomyosin reorganizes, and macromolecule permeability increases in response to inflammation. The dynamic strand model emphasizes the role of claudin cis-interactions and proposes that the transitory breakage of TJ strands regulates the leak pathway. ZO-1 is dynamically connected with TJs, and myosin activity enhances ZO-1 exchange, according to fluorescence recovery after photobleaching investigations (Van Itallie et al. 2017; Shen et al. 2008; Yu et al. 2010). Since ZO proteins are necessary for TJ strand assembly, these findings show that TJ strands are destabilized under inflammatory conditions. They further support the theory that actomyosin-dependent changes in ZO protein dynamics could modulate the leak pathway, which results in TJ strand remodeling. Considering the importance of membrane apposition in the macromolecule permeability barrier, the leak pathway may reflect transient breakage of the trans-interaction sites between neighboring plasma membranes, known as the dynamic membrane apposition model. In this situation, actomyosin contraction induces transient focal widening of the intercellular space and may apply contractile force to TJs, leading to macromolecules leakage along the paracellular space (Otani and Furuse 2020; Mitchell et al. 2015).

6.3.2 Fence Function of Tight Junctions

Besides barrier function, epithelia's other crucial key role involves transporting substances between the external environment and the interior body. Transportation across the epithelium is aided by an electrochemical gradient created by polarized transporter proteins. The electrochemical gradient is generated by epithelial polarity, and the epithelial barrier is required for maintenance. Epithelial polarity and the epithelial barrier must be linked to one another for effective epithelial transport (Cereijido et al. 1998). The asymmetric distribution of membrane proteins exemplifies how TJ assembly is coupled to epithelial polarity establishment (De Camilli et al. 1974; Hoi Sang et al. 1979; Dragsten et al. 1981). Later, it was shown that the membrane lipids in the axoplasmic leaflet could not freely diffuse across TJs, which leads to the idea that TJs can act as a membrane fence that segregates the basolateral and the apical plasma membranes (van Meer and Simons 1986). TJs also play essential roles in epithelial polarity by functioning as scaffolds for polarity signaling proteins, which include Crumbs/Pals1/PATJ complex and Par-3/Par-6/aPKC complex. In comparison, the polarity signaling proteins play an important function in TJ assembly, indicating a reciprocal relationship between TJs and polarity signaling proteins (Otani and Furuse 2020).

6.4 Bidirectional Signaling and Tight Junctions

TJs take part in two main types of signal transduction processes. Signals transmitted from TJs to the cell interior influence gene expression, cell proliferation, and differentiation. Signals from the cell interior to forming or existing TJs guide assembly and modulate their function (Fig. 6.3). The molecular mechanisms that mediate these signal transduction processes are discussed as under:

6.4.1 Signaling to Tight Junctions

TJ assembly and function are regulated by intracellular signaling pathways, which have recently attracted much attention in therapeutics. The activation of TJ assembly and disintegration is caused by a variety of signaling pathways and proteins. Among them are protein kinase C (PKC) isotypes and protein kinase A (PKA), monomeric and heterogenic G proteins (Kojima et al. 2007). Moreover, PKA stimulation prevents TJ disintegration after the calcium removal or antibody-based neutralization of E-cadherin. Additionally, it prevents TJ assembly in a calcium switch protocol, a model system for the de novo formation of intracellular junctions (Balda et al. 1991).

Similarly, PKAs impact on TJ permeability is controversial (Benais-Pont et al. 2003), demonstrating that several factors, including the strength of the stimulus, the cellular background, and interaction with other signaling pathways, may modify the response to PKA stimulation (Balda et al. 1991; de Almeida et al. 1994). Also,

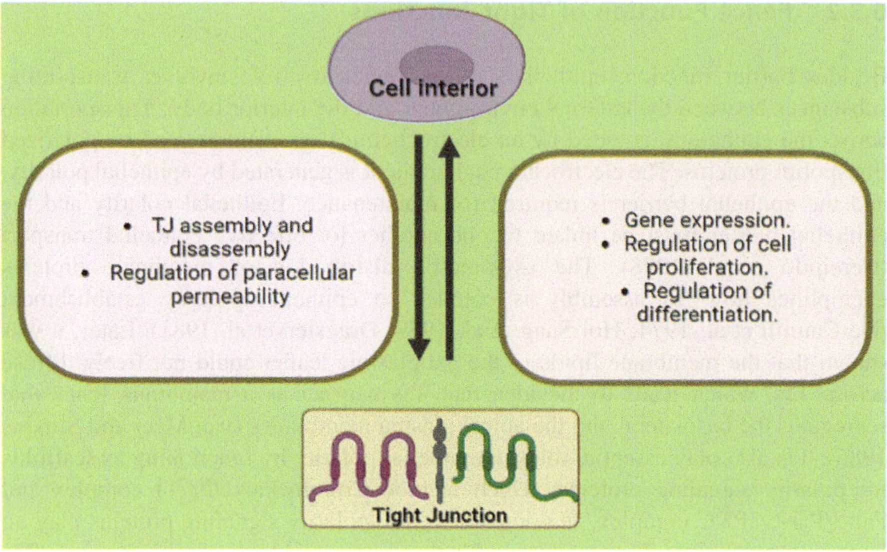

Fig. 6.3 Signal transduction to and from tight junction; systemic representation of the information transmitted from the cell interior to TJs and from TJs to the cell interior. These signal transduction pathways are a complex network of mechanisms that we are just commenced to understand

inhibition of inhibitory G proteins (Gi) stimulates the assembly of TJs, and $G\alpha_{i2}$ was present at cell–cell contacts co-localizing with ZO-1. Whereas ectopically expressed α_i proteins were shown to increase paracellular permeability, which is, in effect, mediated by ZO-1 and stimulates TJ assembly (de Almeida et al. 1994; Meyer et al. 2002). Novel (nPKC), classical (cPKC), and atypical (aPKC) PKCs make up the PKC family of serine/threonine kinases. Phosphorbol ester activates diacylglycerol (DAG), and phosphoserine activates the cPKC (α,β, and γ) and nPKC isotypes. Phosphorbol esters have been demonstrated to cause TJ disintegration in various cell lines as strong activators of the cPKC and nPKC signaling pathways. cPKC also requires calcium for activation, whereas nPKC does not (Mellor and Parker 1998). PAR3 (also known as ASIP, which stands for atypical PKC isotype-specific interaction protein) and PAR6 are PDZ domain proteins found in epithelial TJs. Recruitment to TJs is considered to be mediated when the PAR3 binds to the transmembrane protein JAM. For TJ formation and epithelial polarization, the successful assembly of an active PKC-PAR3-PAR6 complex at TJs is important (Mellor and Parker 1998; Nigam et al. 1991). In addition, TJs have been connected functionally or physically to three types of small GTPases (RAS, Rabs, and Rho family GTPases). The Rho family of GTPases has been related to junction assembly and selective paracellular permeability regulation (Zahraoui et al. 2000). TJ function is disrupted in epithelia and endothelia when constitutively active variants of RhoA and Rac1 are expressed (Jou et al. 1998; Wójciak-Stothard et al. 2001), suggesting their role in TJ disassembly. Rho GTPases that activate the effector proteins lead to the regulation of TJs. Heteromeric G proteins lead to the modulation of TJs and the activation of Rho GTPases (Hasegawa et al. 1999). The activation of Rho by prostaglandins E receptor is regulated by the G protein (Hasegawa et al. 1999). Guanine nucleotide exchange factor (GEF) is generally required for the activation of Rho GTPases, and a TJ-associated GEF for Rho has recently been found. It has been demonstrated that GEF-H1/lfc interacts with TJs and modulates paracellular permeability (Benais-Pont et al. 2003).

6.4.2 Signaling from Tight Junctions

Recently it was observed that components of evolutionary conserved crumbs signaling pathway localize at epithelial TJs. In mammalian epithelial cells, little is known about the functional significance of crumbs and crumbs associated proteins (Tepass et al. 1990). The TJ-associated protein Pals1, which is a stardust homolog, binds to the mammalian homolog discs lost-PATJ and transfected chimeras that contain the cytoplasmic domain of the mammalian crumbs homolog in mammalian epithelial cells (Roh et al. 2002; Lemmers et al. 2002). Overexpression of PATJ interferes with the recruitment of ZO-1 and ZO-3 to form TJs (Adachi et al. 2009). This suggests that crumbs signaling may govern TJ assembly or takes part in TJ-related signaling cascades and controls epithelial proliferation and differentiation (Hurd et al. 2003).

Additionally, the modulation of differentiation is aided by intracellular junctions. In many cancers, the loss of epithelial differentiation often associates with a mutation in small GTPase Ras and is known to mediate extracellular matrix signaling and growth factor and influence cell–cell adhesion (Cary et al. 1999). Afadin AF-6, a junction-associated protein, and other less characterized Ras effectors are also known to contribute to differentiation (Zhadanov et al. 1999). TJs have been associated with two Ras effectors, whose activity triggers the sequential activation of the transcription factor ELK1 and the extracellular signals regulated kinases/mitogen-activated protein kinase (ERK/MAPK) pathway. To transform epithelial cells, expression of RAF1, MAPK, or ERK KINASE 1 (MEK-1) is sufficient (Schramek et al. 1997). TJs seem to be connected intimately with RAF-1 signaling, as overexpression of occludin is enough to reverse RAF-1 mediated transformation of salivary gland epithelial cells (Li and Mrsny 2000). The molecular mechanism by which TJ can counteract RAF signaling has yet to be discovered, and it still needs to be clear whether the inhibition of RAF-1 mediated transformation by occludin involves other signaling molecules or is direct. The PDZ domain protein AF6, which was initially found as a Ras target and localized to TJs, but again which is the rat homolog of AF-6 has since been recruited to nected-based AJs, suggesting that afadin/AF-6 may not be a unique component of TJs. It is not yet clear why the results are disputed because afadin/AF-6 can interact with transmembrane components of both AJs (such as nectin) and TJs such as JAM (Takahashi et al. 1999).

6.5 Deregulation of Tight Junctions and Human Diseases

It is well demonstrated that the deregulation of TJs leads to a marked increment in fence and barrier functions resulting in infection and disease pathogenesis.

6.5.1 Disturbance of the Fence Function

Cancer cells, in general, lose their polarity and many specific functions. RAS protein has been shown to modulate occludin and ZO-1 phosphorylation and thereby suppress TJ activities (Li and Mrsny 2000). In addition, the downregulation of occludin and claudin-1 expression by oncogenic RAF-1 also disrupts TJs that promote phenotypic alterations in malignant cells. Furthermore, a decrease in occludin expression results in a loss of cell polarity in the human colonic glandular epithelium affecting differentiation (Li and Mrsny 2000). Recently, viral oncoprotein c-Kit of human papillomavirus (HPV) and adenovirus has been shown to target RAS substrates, including ALL-1 fusion partner MUPP1.113 AF-6/s-afadin and MAGI proteins involved in acute myeloid leukemia (Glaunsinger et al. 2000; Thomas et al. 2001; Lee et al. 2000). Conversely, claudin-1, -2, -3, and -5 enhance the activation of pro-MMP-2,3,1 (Miyamori et al. 2001), and claudin-4 is overexpressed in ovarian cancer (Tobioka et al. 2002). Except for the tumor suppressor PTEN (Mayo and Donner 2002), relatively few reports have demonstrated genetic alterations in TJ proteins (Hoover et al. 1998). Because of this, claudin dysfunction has been linked to a variety of diseases, including inflammatory bowel disease IBD (Lameris et al. 2013), colorectal cancer (CRC) (Kinugasa et al. 2012), ulcerative colitis (UC) (Kinugasa et al. 2010), and a variety of other malignancies, including breast, gastric, pancreatic, prostate, and uterine cancers. In patients with familial hyperchloremia ichthyosis (Anderson and Van Itallie 2009) and newborn sclerosing cholangitis (NISCH) (Hadj-Rabia et al. 2004) syndrome, mutations in TJ proteins cause abnormalities. These findings could indicate that the loss of TJs in cancer cells is a secondary or late occurrence, even though TJs are thought to be involved in tumorigenesis and metastasis.

6.5.2 Disturbance of the Barrier Function

TJs are implicated in various pathological diseases in which the physiological regulation of ions, molecules, and inflammatory cell passage is disrupted. Table 6.1 compiles the human diseases involving dysfunction of TJs.

The first mechanism is functional alterations of tight junction proteins. The coxsackievirus and adenovirus receptor (CAR) is a multifunctional cellular protein serving as a receptor for adenoviruses, coxsackieviruses, and other viruses besides being involved in cell adhesion, immune cell activation, synaptic transmission, and signaling (Cohen et al. 2001). JAM is a receptor for reovirus and is essential for the reovirus-induced activation of NF-kB (Barton et al. 2001). Claudin- 3 and -4 are

Table 6.1 Human diseases relevant to tight-junction functions

S. No.	Organ affected	Type of the disease	References
Disturbance of the fence function (maintenance of cell polarity)			
1	Cancer	Cancer cells	Li and Mrsny (2000), Tobioka et al. (2002), Chen et al. (2000)
		Oncogenic papillomavirus infection	Thomas et al. (2001), Lee et al. (2000), Tobioka et al. (2002)
Disturbance of barrier function (regulation of paracellular pathway)			
2	Vascular system	Edema	(Marcus et al. 1996)
		Endotoxemia	(Marcus et al. 1996)
		Cytokinemia	(Marcus et al. 1996)
		Diabetic retinopathy	(Antonetti et al. 1999; Murata et al. 1995)
		Multiple sclerosis	(Pardridge 1998)
		Blood-borne metastasis	(Mori et al. 1999; Martin and Jiang 2001)
3	Gastrointestinal tract	Bacterial gastritis	(Papini et al. 1998; Suzuki et al. 2002)
		Pseudomembranous colitis	(Wan et al. 1999)
		Crohn's disease	(Gassler et al. 2001; Kucharzik et al. 2001)
		Ulcerative colitis	(Schmitz et al. 1999)
		Celiac disease	(Fasano et al. 2000)
		Collagenous colitis	(Bürgel et al. 2002)
4	Liver	Jaundice	(Rahner et al. 1996; Takakuwa et al. 2002)
		Primary biliary cirrhosis	(Koizumi et al. 2007)
		Primary sclerosing cholangitis	(Koizumi et al. 2007)
5	Respiratory tract	Asthma	(Wu et al. 2000)
		Adult (acute) respiratory distress syndrome	(Dudek and Garcia 2001)
6	Viral infection	Reovirus, adenovirus, rotavirus	(Barton et al. 2001; Cohen et al. 2001; Svensson et al. 1991)
7	Hereditary disease	Hypomagnesaemia	(Simon et al. 1999)
		Deafness	(Wilcox et al. 2001)
		Cystic fibrosis	(Coyne et al. 2002)

receptors for enterotoxin (CpE) of *Clostridium perfringens* (CPE), a common cause of food poisoning (Katahira et al. 1997). The C-terminal domain of claudins was selectively targeted by CpE (Vecchio et al. 2021). Colonic surface cells that express Claudin-4 bind to CpE and form a cCpE complex, which internalizes like other ligand–receptor complexes, disrupting TJ function (Wan et al. 1999). Dermatophagoides pteronyssinus also digests occludin, allowing it to enter the body and induce asthma by opening TJs in the airway (Wan et al. 1999). V. cholerae's ZOT stimulates PKC, causing TJ loss (Fasano et al. 1995). Also, hemagglutinin/protease (HA/P) produced by *Vibrio cholerae* digests occludin (Wu et al. 2000). Bacteroides fragilis induce TJ breakdown and degrade E-cadherin by its metalloprotease (Wu et al. 1998). By unknown mechanisms, rotavirus and *Helicobacter pylori* vacuolating toxins increase intestinal paracellular permeability (Svensson et al. 1991; Papini et al. 1994). Toxins produced from *Clostridium diphtheria*, *Clostridium difficile*, and enteropathogenic *Escherichia coli* change Rho activity by changing the amino acids to cause severe colitis, the former two causing pseudomembranous colitis (Boquet 1999). Enteropathogenic *E. coli* also induces myosin light chain phosphorylation, which leads to diarrhea (Yuhan et al. 1997). Cytokine production is involved in a wide range of pathological diseases. TNF and interferon downregulate occludins expression, leading to TJ dysfunction (Mankertz et al. 2000); JAM is redistributed in endothelial cells as a result of both causes (Ozaki et al. 1999). Inflammatory bowel illness causes an increase in intestinal permeability, which is caused by occludin downregulation (Gassler et al. 2001; Kucharzik et al. 2001; Schmitz et al. 1999). Zonulin is released when the small intestine is exposed to entero-bacteria, which causes the gut's TJs to open. Zonulin is overexpressed during the acute phase of celiac disease, accompanied by increased intestinal permeability (Fasano et al. 2000; Di Pierro et al. 2001). Inflammation is mainly accompanied by an increase in vascular permeability, which VEGF induces in part, that predominantly impacts the barrier function of TJs by phosphorylating and downregulating occludin (Antonetti et al. 1999; Wang et al. 2001). Cancer cells also release VEGF to promote angiogenesis and, most likely, to intravasate and extravasate or to force their way through the tight connections of the vascular endothelium and spread. Thus, various types of diseases such as metastasis and retinopathy that involve VEGF production involve TJ dysfunction (Antonetti et al. 1999; Satoh et al. 1996). Also, Japanese encephalitis virus (JEV) infection results in change in the blood–brain barrier (BBB) and a drop in claudin-5, ZO1, and occludin was restored by the introduction of neutralizing antibodies against IP-10, an abundant chemokine generated in the early stages of JEV infection, assisting in the reduction of BBB damage. This shows that TJ proteins play a critical function in BBB maintenance (Wang et al. 2018). Hence, it is clear that TJs are targets of pathogens such as viruses, bacteria, and allergens and also change their functions. By the following three mechanisms these pathogens affect the barrier function of TJs (Table 6.2).

6.6 Tight Junctions and Signaling Pathways in Cancer

Over the last decade, TJ has become essential in tumor growth and metastasis. Besides controlling the paracellular transport of ions and certain chemicals, TJ is essential for maintaining tissue integrity and cell–cell adhesion. As in other diseases, the distribution of TJ proteins and changes in the expression promotes the loss of TJ structure cohesiveness associated with the invasive behavior of cancer cells. Because TJs reside between cancer cells and the cells of the endothelium, they are the first structure that obstructs the cancer cells' ability to spread successfully. Against

Table 6.2 Pathogenic agents causing dysfunction of the barrier function

S. No.	Function	Pathogen	References
1	Functional changes of TJ proteins (target molecules)	Clostridium perfringens (Claudin-3,-4)	(Wan et al. 1999; Katahira et al. 1997)
		Vibrio cholera (occludin)	(Wu et al. 2000; Fasano et al. 1995)
		Reovirus (JAM)	(Barton et al. 2001; Cohen et al. 2001)
		Coxsackievirus and adenovirus (CAR)	(Cohen et al. 2001)
		Dermatophagoides pteronyssinus (occluding, claudin-1)	(Wan et al. 1999)
2	Rho and myosin light chain kinase activity alteration results in a change in the organization of actin	*Clostridium diphtheria*	(Boquet 1999)
		Clostridium difficile	(Boquet 1999)
		Enteropathogenic *Escherichia coli*	(Boquet 1999)
3	Indirect mechanisms notably activation of PKC, including unknown ones	Bacteroides fragilis	(Wu et al. 1998)
		Helicobacter pylori	(Svensson et al. 1991; Papini et al. 1994)
		Rotavirus	(Svensson et al. 1991; Papini et al. 1994)

metastatic cancer cells, TJs of vascular endothelium in vivo act as a barrier between blood and tissues. Any change in cancer cells by upregulation or downregulation of significant TJ proteins results in cell contact inhibition and loss of cell–cell association, leading to uncontrolled proliferation, degradation of the basement membrane, and loss of adhesion. It is clear that alterations in the modulation and function of TJ are not just a result of cancer progression. Still, it is important for its development and persistence, ultimately enabling metastasis and secondary disease (Martin 2014).

6.6.1 Crosstalk of Claudins with Signaling Pathways in Cancer

Claudin dysfunction has been linked to various diseases, including IBD and colorectal cancer (CRC), as well as a variety of other cancers involving gastric, breast, pancreatic, uterine, and prostrate (Kinugasa et al. 2012). Claudin-3 and -4 overexpression enhances cancer progression in ovarian cancer cells (Agarwal et al. 2005). Overexpression of these proteins reduces Wnt signaling, E-cadherin expression, and in vitro cell motility and invasion. Downregulation of claudin-3/-4 increases tumor growth and metastasis in ovarian cancer, while in breast cancer, less expression of claudin-3/-4, along with claudin-7, results in high malignancy (Prat et al. 2010). Increased expression of claudin-4, however, suppresses invasion and metastasis in pancreatic cancer cells; similar inhibition is seen in gastric cancer

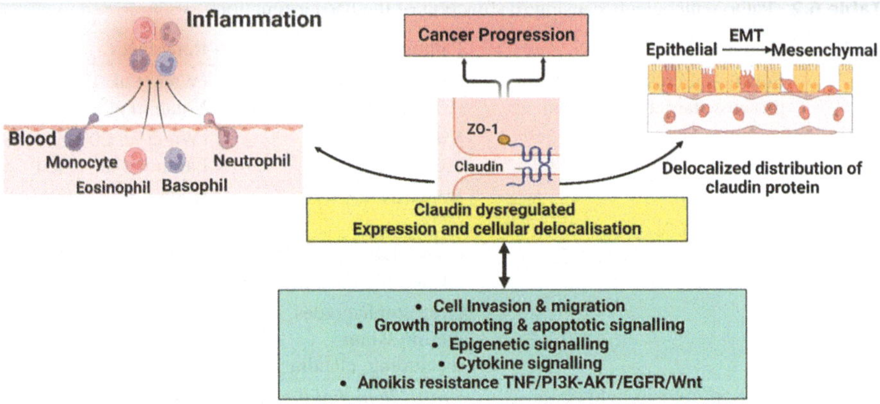

Fig. 6.4 Schematic representation of function and modulation of claudins in the cell during the invasive and metastatic stage. Epigenetic factors, growth factors, and cytokines all contribute to the deregulation of claudin expression and delocalization, which results in the loss of "gate and barrier" function and promotes EMT, inflammation, and disease progression

cells without affecting cell growth (Michl et al. 2003). Low expression of claudin-6 promotes invasiveness in breast cancer, whereas, in gastric cancer cells, less expression provokes invasion, migration, and proliferation (Kwon et al. 2011). Interestingly, claudin-7 functions both as a tumor promoter and suppressor. In esophageal squamous cell carcinoma (ESCC), it enhances cell growth and metastasis, while in colorectal cancer and ovarian cancer, its overexpression leads tumor development and invasiveness (Lioni et al. 2007; Johnson et al. 2005). Though claudins must be phosphorylated to maintain their function, aberrant phosphorylation alters their aggregation and structural stability, potentially compromising epithelial barrier function. The phosphorylation of claudin-1 by mitogen-activated protein kinase (MAPK) (Fujibe et al. 2004), or protein kinase C (PKC) (Ishizaki et al. 2003), and the phosphorylation of claudin-5 by cyclic AMP (cAMP) promote TJ barrier function, whereas protein kinase A leads phosphorylation of claudin-6 and facilitates Mg^{2+} transport (Ikari et al. 2008) (Fig. 6.4).

6.6.1.1 Tight Junctions and Apoptotic/Integrin Signaling Pathways

Anoikis resistance leads to anchorage-independent growth and EMT, both essential for cancer progression and metastatic colonization (Ishizaki et al. 2003). Anoikis resistance of tumor cells involves many mechanisms, with integrin overactivation of receptors and appropriate alterations in the tumor microenvironment (Haenssen et al. 2010). It has been revealed that the signaling molecule Akt plays an important part in anoikis resistance by decreasing pro-apoptotic proteins Bad and caspase-9 by phosphorylation and upregulating the anti-apoptotic protein Bcl2 expression. Furthermore, phosphatidylinositol-3 kinase (PI3K) stimulates Akt, which leads to cell survival in response to integrin-mediated cell attachment (Jeong et al. 2008). Claudin-2 overexpression has been associated with cancer progression in patients with CRC. Claudin-2 overexpression boosted cell proliferation and tumor growth in CRC cells in an EGF receptor (EGFR)-dependent way (Dhawan et al. 2011). Increased claudin-3 expression is associated with increased cell motility and anchorage-independent behavior in human CRC (HT-29) cells, which EGF regulates via the ERK1/2 and PI3K-Akt pathways. Claudin-1 and Src proteins take part in the modulation of anoikis in colon cancer cells via claudin-1/Src/PI3k-Akt/Bcl-2 dependent signaling (de Souza et al. 2013).

6.6.1.2 Tight Junctions and Notch/Wnt Signaling Pathway

Notch signaling can activate or inhibit cellular processes like proliferation, differentiation, and apoptosis, which play a key role in tumorigenesis (Bolós et al. 2007). Signaling *via* Wnt/β-catenin and Notch is necessary for maintaining homeostasis and intestinal development. The Wnt signaling pathway, however, regulates the levels of cytoplasmic and nuclear β-catenin and hence plays a vital role in the development of various tissues and organisms. The inability to coordinate Notch and Wnt signaling has led to enhanced inflammation and tumor progression (Ahmed et al. 2012). Claudin-1 expression increases tumorigenic capacity in CRC, which leads to mucosal inflammation and suppresses goblet cell development by activating the Notch pathway (Bhat et al. 2012). Claudin-1 overexpression was found to regulate Notch signaling, increasing MMP-9 and p-ERK expression in colonic epithelial homeostasis and colon cancer metastasis. Wnt ligands activate the Wnt/β-catenin signaling pathway that modulates embryonic development and homeostasis in later stages. Mislocalization of β-catenin and deregulation of the Wnt/β-catenin signaling pathway have been linked to cancer formation. Because greater than 70% of CRC tumors have mutations in adenomatous polyposis coli (APC), a Wnt pathway component, Wnt/β-catenin signaling becomes more critical (Dhawan et al. 2005). In a metastatic liver lesion, claudin-1 nuclear localization, along with β-catenin, is identified (Dhawan et al. 2005). These numerous significant interactions between TJ proteins and signaling cascades indicate that TJ proteins may have various binding specificities to various signaling molecules and are contextually dependent, which must be examined further.

6.6.1.3 Tight Junctions and Kinase Signaling Pathways

It has been seen that abnormal claudin-1 expression causes significant phenotypic alterations in tumor growth and metastasis. Claudin-1 overexpression boosted metastatic potential by affecting E-cadherin expression and Wnt/β-catenin signaling (Dhawan et al. 2011). Claudin-1 increases MMP activity and promotes invasiveness in oral squamous cell carcinoma and melanomas. Increased claudin-1 expression at both the mRNA and protein levels was associated with PKC activation in human liver cells, which enhances invasiveness by stimulating c-Abl-PKC signaling (Yoon et al. 2010). Claudin-3 and claudin-4 deficiency increased EMT activity in ovarian cancer cells by decreasing E-cadherin expression, increasing twist expression, and activating the PI3K pathway (Lin et al. 2013). Claudin-1 increases EMT in human liver cells, whereas claudin-3 and claudin-4 lead to EMT in ovarian cancer cells, indicating that claudins' effect on EMT is tissue-specific (Yoon et al. 2010).

Claudin-3 and claudin-4 function is regulated by phosphorylation; for example, claudin-3 and claudin-4 are phosphorylated by active PKA or PKC, which promotes paracellular permeability in ovarian cancer cells by leading claudins to be mislocalized. Phosphorylation of claudin-4 by PKC in human pancreatic cancer cells increases its mislocalization and compromises the TJ barrier integrity (Kyuno et al. 2011). The progression of ovarian cancer is associated with overexpression of claudin-3/-4 with concurrent MMP activation, which results in increased invasiveness (Agarwal et al. 2005).

Conversely, it is demonstrated that adherens are responsible for altered claudin-5 expression. Vascular endothelial E-cadherin (VEC) was found to control claudin-5 expression. This was carried out by blocking β-catenin translocation to the nucleus or inactivating FOXO1 inhibitory action via Akt and sequestering it from the nucleus. After treatment with glycogen synthase kinase 3 (GSK-3), the expression of claudin-5 was reduced in VEC-positive cells. Additionally, it was found that FOXO1 and β-catenin are directly associated, and this interaction at the promoter region of claudin-5 is necessary for the regulation and/or overexpression. Without VEC, the FOXO1-β-catenin-Tcf-4 complex links to claudin-5 gene's promoter and prevents its expression (Nunes et al. 2006).

6.6.1.4 Tight Junctions and ERK Signaling Pathway

Phosphorylated by a broad spectrum of signaling pathways, ERK regulates apoptosis, transcription, motility, metabolism, proliferation, cell differentiation, and survival. An increase in ZO-3 expression and a decrease in ZO-2 expression in TJs was associated with a reduction in ERK1/2 phosphorylation (p-ERK) (Kim and Breton 2016). At membrane junctions, it also affects the redistribution of claudin-4 and claudin-1, with no influence on the expression of claudin-3. In TJ integrity, ERK activation plays a contradictory role where its activation causes TJs to be disrupted in some epithelial monolayers while protecting them in other epithelia. This intriguing phenomenon was seen in Caco-2 cell monolayers, where it was found that ERK is involved in the destabilization of TJs in underdifferentiated Caco-2 cells, while a protective role was observed in differentiated cells. The authors hypothesized that this differential effect is due to the variation in the subcellular distribution of ERK (protein phosphatase 2A) with TJ proteins (Aggarwal et al. 2011). TJ proteins have also been demonstrated to activate ERK signaling, which in turn affects cell fate. At the same time, claudin-7 leads to cell growth and metastasis of ESCC (Lioni et al. 2007), although it inhibits migration and invasion of lung cancer cells *via* the ERK/MAPK signaling pathway (Lu et al. 2011). According to another study, the stimulation of the c-Abl-Raf-1/ERK1/2 signaling pathway is primarily responsible for claudin-1's induction of EMT in human liver cells (Suh et al. 2013). This finding supports the significance of c-Abl-ERK signaling in claudin-1-associated malignant phenotype. In the lung cancer A549 cell line, the activation of the EGFR/MEK/ERK signaling pathway was associated with the increased expression of claudin-2 (Ikari et al. 2012). Additionally, it was found that interacting with the AP-1 binding site of the claudin-2 promoter, c-Fos, a downstream target in an EGFR/MEK/ERK pathway, increases the transcriptional activity of claudin-2 (Ikari et al. 2012). In contrast to ERK activation by claudin-2 (Lipschutz et al. 2005), it was shown that the ERK 1/2 signaling pathway is a negative modulator of claudin-2 expression in mammalian renal epithelial cells that affects renal epithelial function and TJ permeability. These studies confirm that claudins were modulated in a tissue-specific manner and that they regulate signaling pathways similarly.

6.7 Conclusion

It can be concluded that TJs have arisen as dynamic bidirectional signaling hubs that hold a variety of regulatory mechanisms for proper junction construction and function. TJ proteins govern cell proliferation, motility, survival, and differentiation by recruiting additional signaling molecules or direct signaling to the cell interior. TJ proteins, particularly claudin family members, are deregulated in various malignancies and inflammatory diseases, making them useful diagnostic and prognostic indicators. The role of claudins in inflammatory illnesses and cancer progression is well understood. The enormous number of pathogenic bacteria and viruses incorporating TJ components are thus of significant interest. To explain how the deregulation of junctional signaling mechanisms contributes to disease development, they offer excellent experimental tools. Therefore, more research in this area is required, and the development of claudin-targeted therapeutics represents a promising endeavor.

Acknowledgment This study was supported by Ramalingaswami Re-entry Fellowship (Grant number: D.O. NO.BT/HRD/35/02/2006) from the Department of Biotechnology, Govt. of India, New Delhi and Core Research Grant (CRG/2021/003805) from Science and Engineering Research Board (SERB), Govt. of India, New Delhi to Muzafar A. Macha and Promotion of University Research and Scientific Excellence (PURSE) Grant to the Islamic University of Science and Technology, Awantipora from Department of Biotechnology, Govt. of India. The authors would also like to acknowledge Sidra Medicine Precision Program for funding and constant support to Ajaz A. Bhat and Ammira S.Al-shabeeb Akil.

Declarations **Ethical Approval and Consent to Participate**
Not applicable.

Consent for Publication Not applicable.

Availability of Supporting Data Not applicable.

Competing Interests The authors declare that they have no competing interests.

References

Adachi M, Hamazaki Y, Kobayashi Y, Itoh M, Tsukita S, Furuse M, Tsukita S (2009) Similar and distinct properties of MUPP1 and Patj, two homologous PDZ domain-containing tight-junction proteins. Mol Cell Biol 29(9):2372–2389

Agarwal R, D'Souza T, Morin PJ (2005) Claudin-3 and claudin-4 expression in ovarian epithelial cells enhances invasion and is associated with increased matrix metalloproteinase-2 activity. Cancer Res 65(16):7378–7385

Aggarwal S, Suzuki T, Taylor WL, Bhargava A, Rao RK (2011) Contrasting effects of ERK on tight junction integrity in differentiated and under-differentiated Caco-2 cell monolayers. Biochem J 433(1):51–63

Ahmed I, Chandrakesan P, Tawfik O, Xia L, Anant S, Umar S (2012) Critical roles of Notch and Wnt/β-catenin pathways in the regulation of hyperplasia and/or colitis in response to bacterial infection. Infect Immun 80(9):3107–3121

Al-Sadi R, Khatib K, Guo S, Ye D, Youssef M, Ma T (2011) Occludin regulates macromolecule flux across the intestinal epithelial tight junction barrier. Am J Physiol Gastrointest Liver Physiol 300(6):1054–1064

Amasheh S, Meiri N, Gitter AH, Schöneberg T, Mankertz J, Schulzke JD, Fromm M (2002) Claudin-2 expression induces cation-selective channels in tight junctions of epithelial cells. J Cell Sci 115(Pt 24):4969–4976

Anderson JM, Van Itallie CM (2009) Physiology and function of the tight junction. Cold Spring Harb Perspect Biol 1(2):a002584

Ando-Akatsuka Y, Yonemura S, Itoh M, Furuse M, Tsukita S (1999) Differential behavior of E-cadherin and occludin in their colocalization with ZO-1 during the establishment of epithelial cell polarity. J Cell Physiol 179(2):115–125

Antonetti DA, Barber AJ, Hollinger LA, Wolpert EB, Gardner TW (1999) Vascular endothelial growth factor induces rapid phosphorylation of tight junction proteins occludin and zonula occluden 1. A potential mechanism for vascular permeability in diabetic retinopathy and tumors. J Biol Chem 274(33):23463–23467

Balda MS, Matter K (2009) Tight junctions and the regulation of gene expression. Biochim Biophys Acta 1788(4):761–767

Balda MS, González-Mariscal L, Contreras RG, Macias-Silva M, Torres-Marquez ME, García-Sáinz JA, Cereijido M (1991) Assembly and sealing of tight junctions: possible participation of G-proteins, phospholipase C, protein kinase C and calmodulin. J Membr Biol 122(3):193–202

Balda MS, Whitney JA, Flores C, González S, Cereijido M, Matter K (1996) Functional dissociation of paracellular permeability and transepithelial electrical resistance and disruption of the apical-basolateral intramembrane diffusion barrier by expression of a mutant tight junction membrane protein. J Cell Biol 134(4):1031–1049

Barton ES, Forrest JC, Connolly JL, Chappell JD, Liu Y, Schnell FJ, Nusrat A, Parkos CA, Dermody TS (2001) Junction adhesion molecule is a receptor for reovirus. Cell 104(3):441–451

Bazzoni G, Martinez-Estrada OM, Orsenigo F, Cordenonsi M, Citi S, Dejana E (2000) Interaction of junctional adhesion molecule with the tight junction components ZO-1, cingulin, and occludin. J Biol Chem 275(27):20520–20526

Benais-Pont G, Punn A, Flores-Maldonado C, Eckert J, Raposo G, Fleming TP, Cereijido M, Balda MS, Matter K (2003) Identification of a tight junction-associated guanine nucleotide exchange factor that activates Rho and regulates paracellular permeability. J Cell Biol 160(5):729–740

Beutel O, Maraspini R, Pombo-García K, Martin-Lemaitre C, Honigmann A (2019) Phase separation of zonula occludens proteins drives formation of tight junctions. Cell 179(4):923–936.e911

Bhat AA, Uppada S, Achkar IW, Hashem S, Yadav SK, Shanmugakonar M, Al-Naemi HA, Haris M, Uddin S (1942) Tight junction proteins and signaling pathways in cancer and inflammation: a functional crosstalk. Front Physiol 2018:9

Bhat AA, Sharma A, Pope J, Krishnan M, Washington MK, Singh AB, Dhawan P (2012) Caudal homeobox protein Cdx-2 cooperates with Wnt pathway to regulate claudin-1 expression in colon cancer cells. PLoS One 7(6):e37174

Bolós V, Grego-Bessa J, de la Pompa JL (2007) Notch signaling in development and cancer. Endocr Rev 28(3):339–363

Boquet P (1999) Bacterial toxins inhibiting or activating small GTP-binding proteins. Ann N Y Acad Sci 886:83–90

Bunt-Milam AH, Saari JC, Klock IB, Garwin GG (1985) Zonulae adherentes pore size in the external limiting membrane of the rabbit retina. Invest Ophthalmol Vis Sci 26(10):1377–1380

Bürgel N, Bojarski C, Mankertz J, Zeitz M, Fromm M, Schulzke JD (2002) Mechanisms of diarrhea in collagenous colitis. Gastroenterology 123(2):433–443

Cary LA, Han DC, Guan JL (1999) Integrin-mediated signal transduction pathways. Histol Histopathol 14(3):1001–1009

Cereijido M, Valdés J, Shoshani L, Contreras RG (1998) Role of tight junctions in establishing and maintaining cell polarity. Annu Rev Physiol 60:161–177

Cereijido M, Shoshani L, Contreras RG (2000) Molecular physiology and pathophysiology of tight junctions. I. Biogenesis of tight junctions and epithelial polarity. Am J Physiol Gastrointest Liver Physiol 279(3):G477–G482

Chattopadhyay N, Wang Z, Ashman LK, Brady-Kalnay SM, Kreidberg JA (2003) alpha3beta1 integrin-CD151, a component of the cadherin-catenin complex, regulates PTPmu expression and cell-cell adhesion. J Cell Biol 163(6):1351–1362

Chen Y, Lu Q, Schneeberger EE, Goodenough DA (2000) Restoration of tight junction structure and barrier function by down-regulation of the mitogen-activated protein kinase pathway in ras-transformed Madin-Darby canine kidney cells. Mol Biol Cell 11(3):849–862

Cohen CJ, Shieh JT, Pickles RJ, Okegawa T, Hsieh JT, Bergelson JM (2001) The coxsackievirus and adenovirus receptor is a transmembrane component of the tight junction. Proc Natl Acad Sci U S A 98(26):15191–15196

Colegio OR, Van Itallie CM, McCrea HJ, Rahner C, Anderson JM (2002) Claudins create charge-selective channels in the paracellular pathway between epithelial cells. Am J Physiol Cell Physiol 283(1):142–147

Cording J, Berg J, Käding N, Bellmann C, Tscheik C, Westphal JK, Milatz S, Günzel D, Wolburg H, Piontek J et al (2013) In tight junctions, claudins regulate the interactions between occludin, tricellulin and marvelD3, which, inversely, modulate claudin oligomerization. J Cell Sci 126(Pt 2):554–564

Coyne CB, Vanhook MK, Gambling TM, Carson JL, Boucher RC, Johnson LG (2002) Regulation of airway tight junctions by proinflammatory cytokines. Mol Biol Cell 13(9):3218–3234

Dbouk HA, Mroue RM, El-Sabban ME, Talhouk RS (2009) Connexins: a myriad of functions extending beyond assembly of gap junction channels. Cell Commun Signal 7:4

de Almeida JB, Holtzman EJ, Peters P, Ercolani L, Ausiello DA, Stow JL (1994) Targeting of chimeric G alpha i proteins to specific membrane domains. J Cell Sci 107(Pt 3):507–515

De Camilli P, Peluchetti D, Meldolesi J (1974) Structural difference between luminal and lateral plasmalemma in pancreatic acinar cells. Nature 248(445):245–247

de Souza WF, Fortunato-Miranda N, Robbs BK, de Araujo WM, de Freitas-Junior JC, Bastos LG, Viola JP, Morgado-Díaz JA (2013) Claudin-3 overexpression increases the malignant potential of colorectal cancer cells: roles of ERK1/2 and PI3K-Akt as modulators of EGFR signaling. PLoS One 8(9):e74994

Dejana E (2004) Endothelial cell-cell junctions: happy together. Nat Rev Mol Cell Biol 5(4): 261–270

Dejana E, Tournier-Lasserve E, Weinstein BM (2009) The control of vascular integrity by endothelial cell junctions: molecular basis and pathological implications. Dev Cell 16(2):209–221

Dhawan P, Singh AB, Deane NG, No Y, Shiou SR, Schmidt C, Neff J, Washington MK, Beauchamp RD (2005) Claudin-1 regulates cellular transformation and metastatic behavior in colon cancer. J Clin Invest 115(7):1765–1776

Dhawan P, Ahmad R, Chaturvedi R, Smith JJ, Midha R, Mittal MK, Krishnan M, Chen X, Eschrich S, Yeatman TJ et al (2011) Claudin-2 expression increases tumorigenicity of colon cancer cells: role of epidermal growth factor receptor activation. Oncogene 30(29):3234–3247

Di Pierro M, Lu R, Uzzau S, Wang W, Margaretten K, Pazzani C, Maimone F, Fasano A (2001) Zonula occludens toxin structure-function analysis. Identification of the fragment biologically active on tight junctions and of the zonulin receptor binding domain. J Biol Chem 276(22): 19160–19165

Dragsten PR, Blumenthal R, Handler JS (1981) Membrane asymmetry in epithelia: is the tight junction a barrier to diffusion in the plasma membrane? Nature 294(5843):718–722
Dudek SM, Garcia JG (2001) Cytoskeletal regulation of pulmonary vascular permeability. J Appl Physiol 91(4):1487–1500
Ebnet K, Schulz CU, Zu M, Brickwedde MK, Pendl GG, Vestweber D (2000) Junctional adhesion molecule interacts with the PDZ domain-containing proteins AF-6 and ZO-1. J Biol Chem 275(36):27979–27988
Farquhar MG, Palade GE (1963) Junctional complexes in various epithelia. J Cell Biol 17(2): 375–412
Fasano A, Fiorentini C, Donelli G, Uzzau S, Kaper JB, Margaretten K, Ding X, Guandalini S, Comstock L, Goldblum SE (1995) Zonula occludens toxin modulates tight junctions through protein kinase C-dependent actin reorganization, in vitro. J Clin Invest 96(2):710–720
Fasano A, Not T, Wang W, Uzzau S, Berti I, Tommasini A, Goldblum SE (2000) Zonulin, a newly discovered modulator of intestinal permeability, and its expression in coeliac disease. Lancet 355(9214):1518–1519
Fujibe M, Chiba H, Kojima T, Soma T, Wada T, Yamashita T, Sawada N (2004) Thr203 of claudin-1, a putative phosphorylation site for MAP kinase, is required to promote the barrier function of tight junctions. Exp Cell Res 295(1):36–47
Fujisawa H, Morioka H, Nakamura H, Watanabe K (1976) Gap junctions in the differentiated neural retinae of newly hatched chickens. J Cell Sci 22(3):597–606
Furuse M, Itoh M, Hirase T, Nagafuchi A, Yonemura S, Tsukita S, Tsukita S (1994) Direct association of occludin with ZO-1 and its possible involvement in the localization of occludin at tight junctions. J Cell Biol 127(6 Pt 1):1617–1626
Furuse M, Furuse K, Sasaki H, Tsukita S (2001) Conversion of zonulae occludentes from tight to leaky strand type by introducing claudin-2 into Madin-Darby canine kidney I cells. J Cell Biol 153(2):263–272
Gassler N, Rohr C, Schneider A, Kartenbeck J, Bach A, Obermüller N, Otto HF, Autschbach F (2001) Inflammatory bowel disease is associated with changes of enterocytic junctions. Am J Physiol Gastrointest Liver Physiol 281(1):G216–G228
Glaunsinger BA, Lee SS, Thomas M, Banks L, Javier R (2000) Interactions of the PDZ-protein MAGI-1 with adenovirus E4-ORF1 and high-risk papillomavirus E6 oncoproteins. Oncogene 19(46):5270–5280
González-Mariscal L, Betanzos A, Nava P, Jaramillo BE (2003) Tight junction proteins. Prog Biophys Mol Biol 81(1):1–44
Günzel D, Fromm M (2012) Claudins and other tight junction proteins. Compr Physiol 2(3): 1819–1852
Hadj-Rabia S, Baala L, Vabres P, Hamel-Teillac D, Jacquemin E, Fabre M, Lyonnet S, De Prost Y, Munnich A, Hadchouel M et al (2004) Claudin-1 gene mutations in neonatal sclerosing cholangitis associated with ichthyosis: a tight junction disease. Gastroenterology 127(5): 1386–1390
Haenssen KK, Caldwell SA, Shahriari KS, Jackson SR, Whelan KA, Klein-Szanto AJ, Reginato MJ (2010) ErbB2 requires integrin alpha5 for anoikis resistance via Src regulation of receptor activity in human mammary epithelial cells. J Cell Sci 123(Pt 8):1373–1382
Hasegawa H, Fujita H, Katoh H, Aoki J, Nakamura K, Ichikawa A, Negishi M (1999) Opposite regulation of transepithelial electrical resistance and paracellular permeability by Rho in Madin-Darby canine kidney cells. J Biol Chem 274(30):20982–20988
Hoi Sang U, Saier MH, Ellisman MH (1979) Tight junction formation is closely linked to the polar redistribution of intramembranous particles in aggregating MDCK epithelia. Exp Cell Res 122(2):384–391
Hoover KB, Liao SY, Bryant PJ (1998) Loss of the tight junction MAGUK ZO-1 in breast cancer: relationship to glandular differentiation and loss of heterozygosity. Am J Pathol 153(6): 1767–1773

Hurd TW, Gao L, Roh MH, Macara IG, Margolis B (2003) Direct interaction of two polarity complexes implicated in epithelial tight junction assembly. Nat Cell Biol 5(2):137–142

Ikari A, Ito M, Okude C, Sawada H, Harada H, Degawa M, Sakai H, Takahashi T, Sugatani J, Miwa M (2008) Claudin-16 is directly phosphorylated by protein kinase A independently of a vasodilator-stimulated phosphoprotein-mediated pathway. J Cell Physiol 214(1):221–229

Ikari A, Sato T, Watanabe R, Yamazaki Y, Sugatani J (2012) Increase in claudin-2 expression by an EGFR/MEK/ERK/c-Fos pathway in lung adenocarcinoma A549 cells. Biochim Biophys Acta 1823(6):1110–1118

Ishizaki T, Chiba H, Kojima T, Fujibe M, Soma T, Miyajima H, Nagasawa K, Wada I, Sawada N (2003) Cyclic AMP induces phosphorylation of claudin-5 immunoprecipitates and expression of claudin-5 gene in blood-brain-barrier endothelial cells via protein kinase A-dependent and -independent pathways. Exp Cell Res 290(2):275–288

Itoh M, Nagafuchi A, Moroi S, Tsukita S (1997) Involvement of ZO-1 in cadherin-based cell adhesion through its direct binding to alpha catenin and actin filaments. J Cell Biol 138(1): 181–192

Itoh M, Furuse M, Morita K, Kubota K, Saitou M, Tsukita S (1999) Direct binding of three tight junction-associated MAGUKs, ZO-1, ZO-2, and ZO-3, with the COOH termini of claudins. J Cell Biol 147(6):1351–1363

Itoh M, Tsukita S, Yamazaki Y, Sugimoto H (2012) Rho GTP exchange factor ARHGEF11 regulates the integrity of epithelial junctions by connecting ZO-1 and RhoA-myosin II signaling. Proc Natl Acad Sci U S A 109(25):9905–9910

Jeong SJ, Dasgupta A, Jung KJ, Um JH, Burke A, Park HU, Brady JN (2008) PI3K/AKT inhibition induces caspase-dependent apoptosis in HTLV-1-transformed cells. Virology 370(2):264–272

Johnson AH, Frierson HF, Zaika A, Powell SM, Roche J, Crowe S, Moskaluk CA, El-Rifai W (2005) Expression of tight-junction protein claudin-7 is an early event in gastric tumorigenesis. Am J Pathol 167(2):577–584

Jou TS, Schneeberger EE, Nelson WJ (1998) Structural and functional regulation of tight junctions by RhoA and Rac1 small GTPases. J Cell Biol 142(1):101–115

Kandouz M, Batist G (2010) Gap junctions and connexins as therapeutic targets in cancer. Expert Opin Ther Targets 14(7):681–692

Katahira J, Sugiyama H, Inoue N, Horiguchi Y, Matsuda M, Sugimoto N (1997) Clostridium perfringens enterotoxin utilizes two structurally related membrane proteins as functional receptors in vivo. J Biol Chem 272(42):26652–26658

Kim B, Breton S (2016) The MAPK/ERK-signaling pathway regulates the expression and distribution of tight junction proteins in the mouse proximal epididymis. Biol Reprod 94(1):22

Kinugasa T, Akagi Y, Yoshida T, Ryu Y, Shiratuchi I, Ishibashi N, Shirouzu K (2010) Increased claudin-1 protein expression contributes to tumorigenesis in ulcerative colitis-associated colorectal cancer. Anticancer Res 30(8):3181–3186

Kinugasa T, Akagi Y, Ochi T, Tanaka N, Kawahara A, Ishibashi Y, Gotanda Y, Yamaguchi K, Shiratuchi I, Oka Y et al (2012) Increased claudin-1 protein expression in hepatic metastatic lesions of colorectal cancer. Anticancer Res 32(6):2309–2314

Koizumi J, Kojima T, Kamekura R, Kurose M, Harimaya A, Murata M, Osanai M, Chiba H, Himi T, Sawada N (2007) Changes of gap and tight junctions during differentiation of human nasal epithelial cells using primary human nasal epithelial cells and primary human nasal fibroblast cells in a noncontact coculture system. J Membr Biol 218(1-3):1–7

Kojima T, Murata M, Go M, Spray DC, Sawada N (2007) Connexins induce and maintain tight junctions in epithelial cells. J Membr Biol 217(1-3):13–19

Kottke MD, Delva E, Kowalczyk AP (2006) The desmosome: cell science lessons from human diseases. J Cell Sci 119(Pt 5):797–806

Krug SM, Amasheh S, Richter JF, Milatz S, Günzel D, Westphal JK, Huber O, Schulzke JD, Fromm M (2009) Tricellulin forms a barrier to macromolecules in tricellular tight junctions without affecting ion permeability. Mol Biol Cell 20(16):3713–3724

Kucharzik T, Walsh SV, Chen J, Parkos CA, Nusrat A (2001) Neutrophil transmigration in inflammatory bowel disease is associated with differential expression of epithelial intercellular junction proteins. Am J Pathol 159(6):2001–2009

Kwon MJ, Kim SH, Jeong HM, Jung HS, Kim SS, Lee JE, Gye MC, Erkin OC, Koh SS, Choi YL et al (2011) Claudin-4 overexpression is associated with epigenetic derepression in gastric carcinoma. Lab Investig 91(11):1652–1667

Kyuno D, Kojima T, Ito T, Yamaguchi H, Tsujiwaki M, Takasawa A, Murata M, Tanaka S, Hirata K, Sawada N (2011) Protein kinase Cα inhibitor enhances the sensitivity of human pancreatic cancer HPAC cells to Clostridium perfringens enterotoxin via claudin-4. Cell Tissue Res 346(3):369–381

Lameris AL, Huybers S, Kaukinen K, Mäkelä TH, Bindels RJ, Hoenderop JG, Nevalainen PI (2013) Expression profiling of claudins in the human gastrointestinal tract in health and during inflammatory bowel disease. Scand J Gastroenterol 48(1):58–69

Lee SS, Glaunsinger B, Mantovani F, Banks L, Javier RT (2000) Multi-PDZ domain protein MUPP1 is a cellular target for both adenovirus E4-ORF1 and high-risk papillomavirus type 18 E6 oncoproteins. J Virol 74(20):9680–9693

Lemmers C, Médina E, Delgrossi MH, Michel D, Arsanto JP, Le Bivic A (2002) hINADl/PATJ, a homolog of discs lost, interacts with crumbs and localizes to tight junctions in human epithelial cells. J Biol Chem 277(28):25408–25415

Li D, Mrsny RJ (2000) Oncogenic Raf-1 disrupts epithelial tight junctions via downregulation of occludin. J Cell Biol 148(4):791–800

Lin X, Shang X, Manorek G, Howell SB (2013) Regulation of the epithelial-mesenchymal transition by claudin-3 and claudin-4. PLoS One 8(6):e67496

Lioni M, Brafford P, Andl C, Rustgi A, El-Deiry W, Herlyn M, Smalley KS (2007) Dysregulation of claudin-7 leads to loss of E-cadherin expression and the increased invasion of esophageal squamous cell carcinoma cells. Am J Pathol 170(2):709–721

Lipschutz JH, Li S, Arisco A, Balkovetz DF (2005) Extracellular signal-regulated kinases 1/2 control claudin-2 expression in Madin-Darby canine kidney strain I and II cells. J Biol Chem 280(5):3780–3788

Lu Z, Ding L, Hong H, Hoggard J, Lu Q, Chen YH (2011) Claudin-7 inhibits human lung cancer cell migration and invasion through ERK/MAPK signaling pathway. Exp Cell Res 317(13): 1935–1946

Mankertz J, Tavalali S, Schmitz H, Mankertz A, Riecken EO, Fromm M, Schulzke JD (2000) Expression from the human occludin promoter is affected by tumor necrosis factor alpha and interferon gamma. J Cell Sci 113(Pt 11):2085–2090

Marcus BC, Wyble CW, Hynes KL, Gewertz BL (1996) Cytokine-induced increases in endothelial permeability occur after adhesion molecule expression. Surgery 120(2):411–416; discussion 416-417

Martin TA (2014) The role of tight junctions in cancer metastasis. Semin Cell Dev Biol 36:224–231

Martin TA, Jiang WG (2001) Tight junctions and their role in cancer metastasis. Histol Histopathol 16(4):1183–1195

Martin TA, Jiang WG (2009) Loss of tight junction barrier function and its role in cancer metastasis. Biochim Biophys Acta 1788(4):872–891

Mayo LD, Donner DB (2002) The PTEN, Mdm2, p53 tumor suppressor-oncoprotein network. Trends Biochem Sci 27(9):462–467

Mellor H, Parker PJ (1998) The extended protein kinase C superfamily. Biochem J 332(Pt 2): 281–292

Meng W, Takeichi M (2009) Adherens junction: molecular architecture and regulation. Cold Spring Harb Perspect Biol 1(6):a002899

Meyer TN, Schwesinger C, Denker BM (2002) Zonula occludens-1 is a scaffolding protein for signaling molecules. Galpha(12) directly binds to the Src homology 3 domain and regulates paracellular permeability in epithelial cells. J Biol Chem 277(28):24855–24858

Michl P, Barth C, Buchholz M, Lerch MM, Rolke M, Holzmann KH, Menke A, Fensterer H, Giehl K, Löhr M et al (2003) Claudin-4 expression decreases invasiveness and metastatic potential of pancreatic cancer. Cancer Res 63(19):6265–6271

Mitchell LA, Ward C, Kwon M, Mitchell PO, Quintero DA, Nusrat A, Parkos CA, Koval M (2015) Junctional adhesion molecule A promotes epithelial tight junction assembly to augment lung barrier function. Am J Pathol 185(2):372–386

Miyamori H, Takino T, Kobayashi Y, Tokai H, Itoh Y, Seiki M, Sato H (2001) Claudin promotes activation of pro-matrix metalloproteinase-2 mediated by membrane-type matrix metalloproteinases. J Biol Chem 276(30):28204–28211

Mori M, Sawada N, Kokai Y, Satoh M (1999) Role of tight junctions in the occurrence of cancer invasion and metastasis. Med Electron Microsc 32(4):193–198

Morris MA, Young LS, Dawson CW (2008) DNA tumour viruses promote tumour cell invasion and metastasis by deregulating the normal processes of cell adhesion and motility. Eur J Cell Biol 87(8-9):677–697

Murata T, Ishibashi T, Khalil A, Hata Y, Yoshikawa H, Inomata H (1995) Vascular endothelial growth factor plays a role in hyperpermeability of diabetic retinal vessels. Ophthalmic Res 27(1):48–52

Nigam SK, Denisenko N, Rodriguez-Boulan E, Citi S (1991) The role of phosphorylation in development of tight junctions in cultured renal epithelial (MDCK) cells. Biochem Biophys Res Commun 181(2):548–553

Nunes FD, Lopez LN, Lin HW, Davies C, Azevedo RB, Gow A, Kachar B (2006) Distinct subdomain organization and molecular composition of a tight junction with adherens junction features. J Cell Sci 119(Pt 23):4819–4827

Ooshio T, Kobayashi R, Ikeda W, Miyata M, Fukumoto Y, Matsuzawa N, Ogita H, Takai Y (2010) Involvement of the interaction of afadin with ZO-1 in the formation of tight junctions in Madin-Darby canine kidney cells. J Biol Chem 285(7):5003–5012

Otani T, Furuse M (2020) Tight junction structure and function revisited. Trends Cell Biol 30(10): 805–817

Otani T, Nguyen TP, Tokuda S, Sugihara K, Sugawara T, Furuse K, Miura T, Ebnet K, Furuse M (2019) Claudins and JAM-A coordinately regulate tight junction formation and epithelial polarity. J Cell Biol 218(10):3372–3396

Ozaki H, Ishii K, Horiuchi H, Arai H, Kawamoto T, Okawa K, Iwamatsu A, Kita T (1999) Cutting edge: combined treatment of TNF-alpha and IFN-gamma causes redistribution of junctional adhesion molecule in human endothelial cells. J Immunol 163(2):553–557

Papini E, de Bernard M, Milia E, Bugnoli M, Zerial M, Rappuoli R, Montecucco C (1994) Cellular vacuoles induced by Helicobacter pylori originate from late endosomal compartments. Proc Natl Acad Sci U S A 91(21):9720–9724

Papini E, Satin B, Norais N, de Bernard M, Telford JL, Rappuoli R, Montecucco C (1998) Selective increase of the permeability of polarized epithelial cell monolayers by Helicobacter pylori vacuolating toxin. J Clin Invest 102(4):813–820

Pardridge WM (1998) Blood-brain barrier carrier-mediated transport and brain metabolism of amino acids. Neurochem Res 23(5):635–644

Prat A, Parker JS, Karginova O, Fan C, Livasy C, Herschkowitz JI, He X, Perou CM (2010) Phenotypic and molecular characterization of the claudin-low intrinsic subtype of breast cancer. Breast Cancer Res 12(5):R68

Rahner C, Stieger B, Landmann L (1996) Structure-function correlation of tight junctional impairment after intrahepatic and extrahepatic cholestasis in rat liver. Gastroenterology 110(5):1564–1578

Richter JF, Hildner M, Schmauder R, Turner JR, Schumann M, Reiche J (2019) Occludin knockdown is not sufficient to induce transepithelial macromolecule passage. Tissue Barriers 7(2):1612661

Roh MH, Makarova O, Liu CJ, Shin K, Lee S, Laurinec S, Goyal M, Wiggins R, Margolis B (2002) The Maguk protein, Pals1, functions as an adapter, linking mammalian homologues of Crumbs and Discs Lost. J Cell Biol 157(1):161–172

Satoh H, Zhong Y, Isomura H, Saitoh M, Enomoto K, Sawada N, Mori M (1996) Localization of 7H6 tight junction-associated antigen along the cell border of vascular endothelial cells correlates with paracellular barrier function against ions, large molecules, and cancer cells. Exp Cell Res 222(2):269–274

Sawada N, Murata M, Kikuchi K, Osanai M, Tobioka H, Kojima T, Chiba H (2003) Tight junctions and human diseases. Med Electron Microsc 36(3):147–156

Schmitz H, Barmeyer C, Fromm M, Runkel N, Foss HD, Bentzel CJ, Riecken EO, Schulzke JD (1999) Altered tight junction structure contributes to the impaired epithelial barrier function in ulcerative colitis. Gastroenterology 116(2):301–309

Schramek H, Feifel E, Healy E, Pollack V (1997) Constitutively active mutant of the mitogen-activated protein kinase kinase MEK1 induces epithelial dedifferentiation and growth inhibition in madin-darby canine kidney-C7 cells. J Biol Chem 272(17):11426–11433

Schwayer C, Shamipour S, Pranjic-Ferscha K, Schauer A, Balda M, Tada M, Matter K, Heisenberg CP (2019) Mechanosensation of tight junctions depends on ZO-1 phase separation and flow. Cell 179(4):937–952.e918

Shen L, Weber CR, Turner JR (2008) The tight junction protein complex undergoes rapid and continuous molecular remodeling at steady state. J Cell Biol 181(4):683–695

Shen L, Weber CR, Raleigh DR, Yu D, Turner JR (2011) Tight junction pore and leak pathways: a dynamic duo. Annu Rev Physiol 73:283–309

Simon DB, Lu Y, Choate KA, Velazquez H, Al-Sabban E, Praga M, Casari G, Bettinelli A, Colussi G, Rodriguez-Soriano J et al (1999) Paracellin-1, a renal tight junction protein required for paracellular Mg2+ resorption. Science 285(5424):103–106

Smith CL, Reese TS (2016) Adherens junctions modulate diffusion between epithelial cells in Trichoplax adhaerens. Biol Bull 231(3):216–224

Steed E, Balda MS, Matter K (2010) Dynamics and functions of tight junctions. Trends Cell Biol 20(3):142–149

Suh Y, Yoon CH, Kim RK, Lim EJ, Oh YS, Hwang SG, An S, Yoon G, Gye MC, Yi JM et al (2013) Claudin-1 induces epithelial-mesenchymal transition through activation of the c-Abl-ERK signaling pathway in human liver cells. Oncogene 32(41):4873–4882

Suzuki K, Kokai Y, Sawada N, Takakuwa R, Kuwahara K, Isogai E, Isogai H, Mori M (2002) SS1 Helicobacter pylori disrupts the paracellular barrier of the gastric mucosa and leads to neutrophilic gastritis in mice. Virchows Arch 440(3):318–324

Suzuki H, Nishizawa T, Tani K, Yamazaki Y, Tamura A, Ishitani R, Dohmae N, Tsukita S, Nureki O, Fujiyoshi Y (2014) Crystal structure of a claudin provides insight into the architecture of tight junctions. Science 344(6181):304–307

Suzuki H, Tani K, Tamura A, Tsukita S, Fujiyoshi Y (2015) Model for the architecture of claudin-based paracellular ion channels through tight junctions. J Mol Biol 427(2):291–297

Svensson L, Finlay BB, Bass D, von Bonsdorff CH, Greenberg HB (1991) Symmetric infection of rotavirus on polarized human intestinal epithelial (Caco-2) cells. J Virol 65(8):4190–4197

Takahashi K, Nakanishi H, Miyahara M, Mandai K, Satoh K, Satoh A, Nishioka H, Aoki J, Nomoto A, Mizoguchi A et al (1999) Nectin/PRR: an immunoglobulin-like cell adhesion molecule recruited to cadherin-based adherens junctions through interaction with Afadin, a PDZ domain-containing protein. J Cell Biol 145(3):539–549

Takakuwa Y, Kokai Y, Sasaki K, Chiba H, Tobioka H, Mori M, Sawada N (2002) Bile canalicular barrier function and expression of tight-junctional molecules in rat hepatocytes during common bile duct ligation. Cell Tissue Res 307(2):181–189

Tepass U, Theres C, Knust E (1990) crumbs encodes an EGF-like protein expressed on apical membranes of Drosophila epithelial cells and required for organization of epithelia. Cell 61(5):787–799

Thomas M, Glaunsinger B, Pim D, Javier R, Banks L (2001) HPV E6 and MAGUK protein interactions: determination of the molecular basis for specific protein recognition and degradation. Oncogene 20(39):5431–5439

Tobioka H, Isomura H, Kokai Y, Sawada N (2002) Polarized distribution of carcinoembryonic antigen is associated with a tight junction molecule in human colorectal adenocarcinoma. J Pathol 198(2):207–212

Tsukita S, Furuse M (2002) Claudin-based barrier in simple and stratified cellular sheets. Curr Opin Cell Biol 14(5):531–536

Tsukita S, Furuse M, Itoh M (2001) Multifunctional strands in tight junctions. Nat Rev Mol Cell Biol 2(4):285–293

Van Itallie CM, Tietgens AJ, Anderson JM (2017) Visualizing the dynamic coupling of claudin strands to the actin cytoskeleton through ZO-1. Mol Biol Cell 28(4):524–534

van Meer G, Simons K (1986) The function of tight junctions in maintaining differences in lipid composition between the apical and the basolateral cell surface domains of MDCK cells. EMBO J 5(7):1455–1464

Vecchio AJ, Rathnayake SS, Stroud RM (2021) Structural basis for Clostridium perfringens enterotoxin targeting of claudins at tight junctions in mammalian gut. Proc Natl Acad Sci U S A 118:15

Wan H, Winton HL, Soeller C, Tovey ER, Gruenert DC, Thompson PJ, Stewart GA, Taylor GW, Garrod DR, Cannell MB et al (1999) Der p 1 facilitates transepithelial allergen delivery by disruption of tight junctions. J Clin Invest 104(1):123–133

Wang W, Dentler WL, Borchardt RT (2001) VEGF increases BMEC monolayer permeability by affecting occludin expression and tight junction assembly. Am J Phys Heart Circ Phys 280(1): H434–H440

Wang K, Wang H, Lou W, Ma L, Li Y, Zhang N, Wang C, Li F, Awais M, Cao S et al (2018) IP-10 promotes blood-brain barrier damage by inducing tumor necrosis factor alpha production in Japanese encephalitis. Front Immunol 9:1148

Watson CJ, Hoare CJ, Garrod DR, Carlson GL, Warhurst G (2005) Interferon-gamma selectively increases epithelial permeability to large molecules by activating different populations of paracellular pores. J Cell Sci 118(Pt 22):5221–5230

Weber CR, Liang GH, Wang Y, Das S, Shen L, Yu AS, Nelson DJ, Turner JR (2015) Claudin-2-dependent paracellular channels are dynamically gated. elife 4:e09906

Wilcox ER, Burton QL, Naz S, Riazuddin S, Smith TN, Ploplis B, Belyantseva I, Ben-Yosef T, Liburd NA, Morell RJ et al (2001) Mutations in the gene encoding tight junction claudin-14 cause autosomal recessive deafness DFNB29. Cell 104(1):165–172

Wójciak-Stothard B, Potempa S, Eichholtz T, Ridley AJ (2001) Rho and Rac but not Cdc42 regulate endothelial cell permeability. J Cell Sci 114(Pt 7):1343–1355

Wu S, Lim KC, Huang J, Saidi RF, Sears CL (1998) Bacteroides fragilis enterotoxin cleaves the zonula adherens protein, E-cadherin. Proc Natl Acad Sci U S A 95(25):14979–14984

Wu Z, Nybom P, Magnusson KE (2000) Distinct effects of Vibrio cholerae haemagglutinin/protease on the structure and localization of the tight junction-associated proteins occludin and ZO-1. Cell Microbiol 2(1):11–17

Yoon CH, Kim MJ, Park MJ, Park IC, Hwang SG, An S, Choi YH, Yoon G, Lee SJ (2010) Claudin-1 acts through c-Abl-protein kinase Cdelta (PKCdelta) signaling and has a causal role in the acquisition of invasive capacity in human liver cells. J Biol Chem 285(1):226–233

Yu AS, Cheng MH, Angelow S, Günzel D, Kanzawa SA, Schneeberger EE, Fromm M, Coalson RD (2009) Molecular basis for cation selectivity in claudin-2-based paracellular pores: identification of an electrostatic interaction site. J Gen Physiol 133(1):111–127

Yu D, Marchiando AM, Weber CR, Raleigh DR, Wang Y, Shen L, Turner JR (2010) MLCK-dependent exchange and actin binding region-dependent anchoring of ZO-1 regulate tight junction barrier function. Proc Natl Acad Sci U S A 107(18):8237–8241

Yuhan R, Koutsouris A, Savkovic SD, Hecht G (1997) Enteropathogenic Escherichia coli-induced myosin light chain phosphorylation alters intestinal epithelial permeability. Gastroenterology 113(6):1873–1882

Zahraoui A, Louvard D, Galli T (2000) Tight junction, a platform for trafficking and signaling protein complexes. J Cell Biol 151(5):31–36

Zhadanov AB, Provance DW Jr, Speer CA, Coffin JD, Goss D, Blixt JA, Reichert CM, Mercer JA (1999) Absence of the tight junctional protein AF-6 disrupts epithelial cell-cell junctions and cell polarity during mouse development. Curr Biol 9(16):880–888

Zhan T, Rindtorff N, Boutros M (2017) Wnt signaling in cancer. Oncogene 36(11):1461–1473

Zihni C, Mills C, Matter K, Balda MS (2016) Tight junctions: from simple barriers to multifunctional molecular gates. Nat Rev Mol Cell Biol 17(9):564–580

Molecular Architecture and Function of Tight Junctions

7

Mudasir A. Kumar, Tulaib Azam Khan, Sara K. Al Marzooqi, Alanoud Abdulla, Tariq Masoodi, Ammira S. Al-Shabeeb Akil, Ajaz A. Bhat, and Muzafar A. Macha

Abstract

Tight junctions, also called zonula occludens, are supramolecular cell–cell adhesion complexes crucial for the architecture of epithelial tissues and selective gates that regulate the paracellular diffusion of ions and solutes. These tight connections prevent the intermixing of apical and basolateral membrane components by generating intramembrane diffusion barriers. Tight junction complexes of adjacent cells also create paracellular channels because of how closely they are spaced, which enables the selective diffusion of ions and solutes through the extracellular environment. Tight junctions are also connected to various cell behaviors and functions, such as controlling cell growth and differentiation, through the transmission of information to the cytoskeleton, nucleus, and different cell adhesion complexes. In addition to inherent genetic changes, recent studies have shown that bacterial toxins, cytokines, hormones, and drugs modify tight junction protein complexes, thus affecting their cellular functions. Recent studies have broadened our understanding of tight junction molecular

---

Mudasir A. Kumar and Tulaib Azam Khan contributed equally as first authors.

M. A. Kumar · M. A. Macha (✉)
Watson-Crick Centre for Molecular Medicine, Islamic University of Science and Technology, Awantipora, Jammu and Kashmir, India

T. A. Khan
Department of Biotechnology, School of Life Sciences, Central University of Kashmir, Ganderbal, Jammu and Kashmir, India

S. K. Al Marzooqi · A. Abdulla · A. S. A.-S. Akil · A. A. Bhat
Department of Human Genetics-Precision Medicine in Diabetes, Obesity and Cancer Research Program, Sidra Medicine, Doha, Qatar

T. Masoodi
Laboratory of Cancer Immunology and Genetics, Sidra Medicine, Doha, Qatar

A. A. Bhat et al. (eds.), *Tight Junctions in Inflammation and Cancer*,
https://doi.org/10.1007/978-981-99-2415-8_7

architecture, biogenesis, and regulation mechanisms. Here we discuss the architecture and functions of this supramolecular complex and its alteration in gene expression and structural architecture in disease states.

Keywords

Apoptosis · Cancer · Cell adhesion · Cell proliferation · Claudins · Tight junctions · Zonula occludens · Junctional adhesion molecules

Abbreviations

CpE	Clostridium perfringens enterotoxins
CRC	Colorectal cancer
D1Ga	Drosophila disc-large tumor suppressor
ECL	Extracellular loop
ECM	Extracellular matrix
ErbB2	Erythroblastic oncogene B
FFEM	Freeze-fracture electron microscopy
GRHL2	Grainyhead like transcription factor 2
HGF	Hepatocyte growth factor
JAM	Junctional adhesion molecule
MAGI	Membrane associated guanylate kinase inverted
MDCK	Madin-Darby canine kidney
MLC	Myosin regulatory light chain
MPDZ	Multiple PDZ proteins
NISCH	Neonatal ichthyosis-sclerosing cholangitis syndrome
PTEN	Phosphatase and tensin homolog
TAMP	Tight junction associated MARVEL domain-containing proteins
TER	Trans-epithelial resistance
TJs	Tight junctions
TRIC	Tailless complex polypeptide 1 ring complex
TTJ	Tricellular tight junction
ZO	Zona occluding
ZONAB	Zonula occludens 1-associated nucleic acid-binding protein

7.1 Introduction

Cell adhesion is an important property that mediates cells to bind and communicate with each other. The highly dynamic cell adhesion process between neighboring cells with their extracellular matrix (ECM) is sustained by cell surface molecules called receptors. It is required for cell polarization and tissue morphogenesis. Junctions between the cells and ECM are a multiprotein complex structure called cell junctions that controls the transport of substances across the cell layers. There are four different intercellular junctions, and each has peculiar transmembrane and sub-membranous protein plaques connected to the actin cytoskeleton.

(a) Tight junctions (TJs, zonula occludens) is a multiprotein complex that regulates semi-permeable diffusion between individual cells and thus limits the mixing of the components of the apical and basolateral membranes (Anderson and Van Itallie 2009; Shen et al. 2011; Zihni et al. 2016).
(b) Adhesion junctions (zonula adhesions) are a multiprotein architecture that controls cell adhesion.
(c) Desmosomes (macula adherens) are specialized intercellular anchoring junctions connecting adjacent cells.
(d) Gap junctions (macula communicans) allow direct molecular exchange between cells by establishing an intercellular channel.

Tight junctions are the complex supramolecular, evolutionary conserved, and most abundant apical member of the intercellular junction system. These supramolecular complex structures are made up of membrane-assisting proteins, integral membrane proteins, and soluble cytoplasmic proteins, which participate in the complex and active process of protein–protein interactions (Fig. 7.1). The pioneering work and ultrastructure of TJs were proposed by Farquhar et al. over 50 years ago and considered tight junctions as diffusion barriers and seals (Farquhar and Palade 1963). However, studies using freeze-fracture electron microscopy (FFEM) showed that TJs are made up of rows of membrane connections that change in number and shape between various tissues. In recent years, there has been an explosion of information regarding the structure of TJs at the molecular level.

The sealing function of TJs in paracellular spaces allows a directional flow of ions and solutes across the cell layers. Selective paracellular diffusion (based on charge and size) is important for maintaining homeostasis in organs and tissues. Recently, the identification of other protein components of TJs imparting a wide range of functions has challenged the traditional dogma of their role only as barrier proteins. While the mutations in the genes encoding TJ structural complexes have recently been associated with many diseases. These complexes are also an ideal target for various pathogenic attacks by directly either destabilizing the junctional network to enter the cell or affecting signaling pathways. TJs can also be regulated in response to various stimuli, such as nutrients and hormones, which can alter both the permeability of ions and solutes and the functional characteristics of TJs (Wong and Gumbiner 1997; Chen et al. 1997).

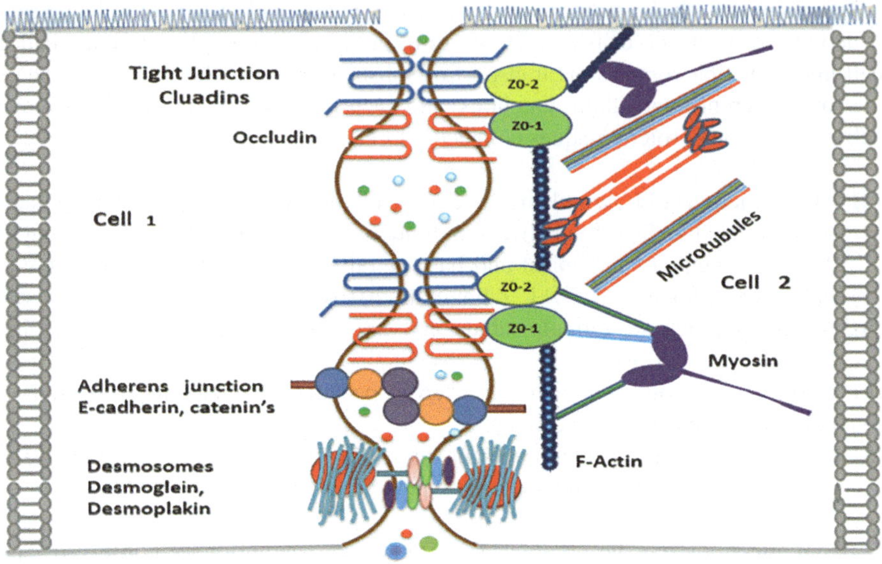

Fig. 7.1 Schematic representation of the different components of the intracellular epithelial junction

7.2 Models of Tight Junction Architecture

Based on the ultrastructural appearance, two different models of the TJ have been put forward, including (a) the protein model and (b) the protein–lipid hybrid model (also known as the lipid model) (Fig. 7.2). The protein model postulates that the TJs are formed by transmembrane proteins forming an intercellular complex between neighboring cell membranes having standard bilayer lipid configuration. The protein model is supported due to the identification of the claudin 15 crystal structure. The 2.4 Å resolution structure of claudin 15 disclosed a characteristic β-sheet fold, comprising two extracellular domains attached to a transmembrane four-helix bundle by a consensus motif. Linear polymers of claudin 15 are existed by specific interactions between adjacent extracellular domains, which are essential for reconstituting TJ-like intramembrane strands. A new model is existed by Cys crosslinking in which two such antiparallel strands associate with each other to form an intramembrane TJ strand. Consequently, four claudin polymers, two per cell, are present at each membrane contact site that interacts through their extracellular domains, forming the junctional barrier and permeation pathway. Polymerization of claudins proteins in the same or adjacent strand in a homotypic or heterotypic manner, TJ strands form a mosaic consisting of various claudin molecules (Furuse et al. 1999; Tsukita and Furuse 1999; Haseloff et al. 2015). The hybrid model proposes that close inter-membrane junctions or TJs are the actual membrane hemifusion, and intramembrane strands are the cylinders of inverted lipid micelles

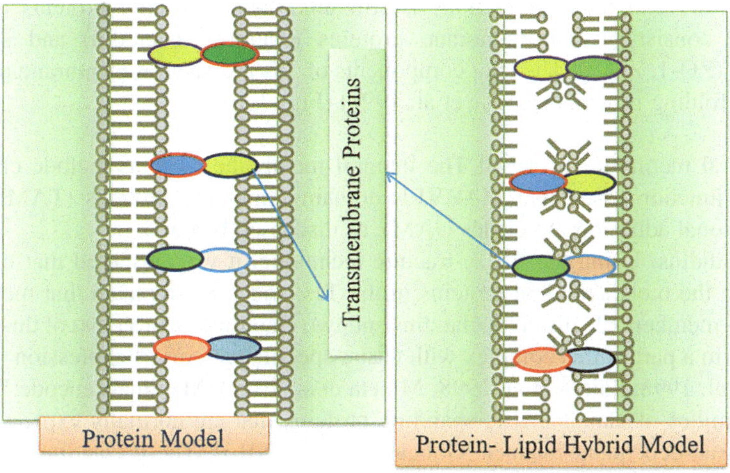

Fig. 7.2 Models of tight junction structure. The protein model suggests that the TJ barrier is created by intercellular protein–protein interactions (ovule-shaped) between neighboring cells. This establishes a paracellular diffusion barrier between two plasma membranes made of common lipid bilayers. The protein–lipid hybrid model proposes that cylinder-shaped inverted micelles supported by transmembrane proteins cause discontinuity of the apical and lateral exoplasmic leaflets, which results in hemifusion areas between the two adjacent plasma membranes. Continuous exoplasmic leaflets exist between neighboring cells in this concept

(Pinto da Silva and Kachar 1982). The basis of this model is the formation of junction-like strands by protein-free liposomes. Though the hemifusion state is energetically unstable, transmembrane proteins are proposed to stabilize the inverted micelle structure (Chernomordik and Kozlov 2005; Kan 1993). This model proposes that lipids cover the spaces between various membrane protein types, eliminating the requirement for proteins to form continuous polymers (Fujimoto 1995; Shen et al. 2008). Some proteins of the protein hybrid model are associated with cholesterol-rich, detergent-resistant membrane micro-domains. Alteration and the reduction of membrane cholesterol content can change the properties of the epithelial barrier property (Cording et al. 2013; Nusrat et al. 2000; Lambert et al. 2005; Lynch et al. 2007).

7.3 Architecture of TJs

In the mid-1980, with the help of immuno-electron microscopy (immune-EM), zona occluding-1 (ZO-1) was the first protein component of the TJ identified (Stevenson et al. 1986; Jesaitis and Goodenough 1994; Haskins et al. 1998). This was followed by identifying two other related proteins, ZO-2 and ZO-3, and an unrelated protein named cingulin. All these proteins were attributed as peripheral proteins. While Taukita et al. first identified a transmembrane proteins in the TJ (Citi et al. 1988; Furuse et al. 1993), complete and exact list of all TJ components and well-defined

3D structure and their interactions is still unknown. The core proteins of tight junction consist of transmembrane proteins (claudins, occludin) and scaffold proteins (ZO-1, ZO-2). The core components of TJs are integral membrane proteins and scaffolding proteins (Furuse et al. 1998) (Fig. 7.3).

1. Integral membrane proteins: The integral membrane proteins include claudins, tight junction associated MARVEL domain-containing proteins (TAMP), and junctional adhesion molecules (JAM), as discussed below.
   a. Claudins: Using the freeze-fracture technique, it was observed that claudins are the most abundant proteins in the TJs, and it is estimated that more than 25 members (23 different claudins and two claudin-like proteins) of this family form a part of this complex with tissue-specific patterns of expression (Furuse et al. 1998; Saitou et al. 1998; Mineta et al. 2011). Mammals encode 27-gene families of claudins tetra-spanning proteins that are normally expressed in a tissue-dependent manner (Gonçalves et al. 2013). These claudins are small molecular weight proteins (20–25 kDa). They are arranged either as extended strands as homophilic or heterophilic within the same membrane (cis-pairing) or across the membranes (trans-pairing). All the claudins share a small region of conserved residues but differ in the length of two different termini. Based on the degree of sequence similarity, claudins are classified into classic (1–10, 14, 15, 17, 19) and non-classic (11–13, 16, 18, 20–24) (Krause et al. 2008). Further, based on functions, they are grouped into (**a**) sealing/barrier claudins (1, 3, 5, 11, 14, 19), (**b**) cationic/anionic/permeable of water (2, 10a, 10b, 15, 17) (Günzel and Fromm 2012), and (**c**) undetermined (4, 7, 8, and 16). These claudins and other cytoplasmic factors regulate processes like permeability of tight junctions, cell signaling, cell cycle regulation, maintenance of cell polarity, and vesicle trafficking (Günzel and Yu 2013).
      In 2014, the high-resolution crystallographic structure of one of the full-length mouse claudin-15 was elucidated and it was observed that a tetrameric channel was formed where four aspartate-15 residues act as a hurdle for anions (Günzel and Yu 2013). This selectively favors the flow of cations over anions. The protein structure is arranged as four antiparallel transmembrane helices with the amino and the carboxy-terminus toward the cytoplasmic side. A long extracellular loop, ECL-1 (strands 1-4), and a short ECL-2 (strand 5, pairing with 1) form five up-and-down antiparallel sheets in the extracellular part (Samanta et al. 2018). It is believed that five up-and-down antiparallel-sheet structures and the other related claudin structures provide the basis for a target site for drug administration (Samanta et al. 2018). Similarly, proline-134 in the transmembrane helix α3 of mice claudin-3 has been shown to induce a bend in the helix. It is worth mentioning that the presence of proline-134 is a peculiar feature of most human claudins. It has recently been shown that ECLs act as a seat of attachment for different toxins like clostridium perfringens enterotoxins (CpE) (Saitoh et al. 2015). ECL of claudin-3 upon binding with CpE results in pore formation in the plasma membrane of mucosa cells of the host and leads to clinical symptoms of CPE intoxication (Fernández Miyakawa et al. 2005).

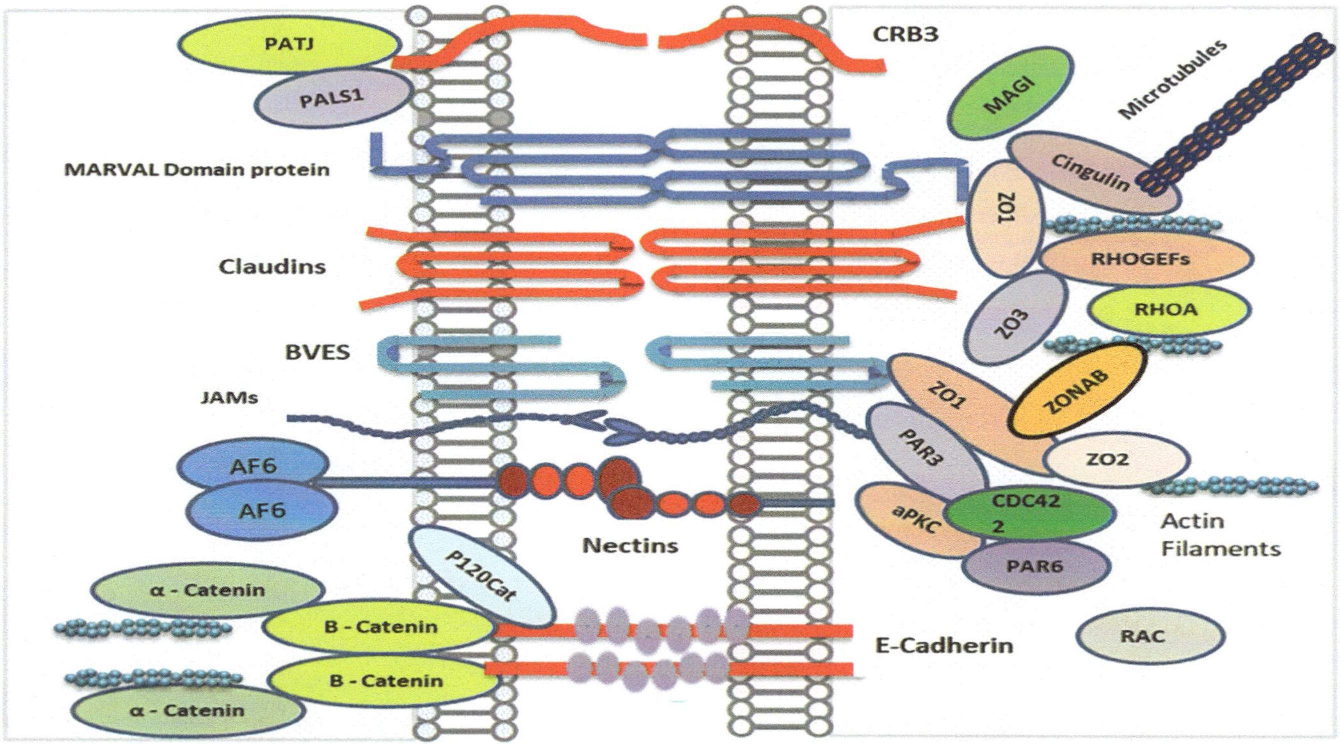

Fig. 7.3 Tight junction core and related proteins; only key group representatives are shown. Blood vessel epicardial substance (BVES), junctional adhesion molecules (JAMs), and other immunoglobulin (Ig)-type adhesion proteins are examples of TM proteins of tight junction (red), which interact with a complex cytoplasmic protein network. The cytoplasmic plaque serves as a physical link between the cytoskeleton and these adhesion proteins (microtubules, actin filaments). Zonula occludens (ZO) proteins ZO1, ZO2, and ZO3 as well as cingulin, membrane-associated guanylate kinase inverted (MAGI), polarity proteins

Fig. 7.3 (continued) such as partitioning defective 3 (PAR3) and PAR6, protein associated with Lin 1 (PALS1), and PALS1-associated tight junction are examples of adaptor proteins and cytoskeletal linkers (PATJ). Signaling components associated with tight junctions include atypical protein kinase C (aPKC); the small RHO GTPases CDC42, RAC and RHOA; and their regulators, guanine nucleotide exchange factors for RHO GTPases (RHOGEFS). Additionally, ZO-1, a submembrane protein of tight junctions, connects to the SH3 domain through ZONAB (ZO-1-associated nucleic acid-binding protein). Nectins and the essential single-span protein E-cadherin, which in the cytoplasm binds with catenin and p120 catenin, also contribute to the formation of tight junctions

This was evident by elucidating the crystal structure of a full-length claudin-19 and its interaction with CpE and subsequent stabilization of v1 loop associated with increased epithelial permeability. Similar interactions of CpE are found with other claudins as well.

b. Tight Junction Associated MARVEL Domain-Containing Proteins
Tight junction associated MARVEL domain-containing proteins (TAMPS) are members of a special family of proteins that possess tetra-span transmembrane domains and share a likeness with myelin and lymphocyte-associated proteins. Occludin and tricellulin are the two members of TAMPs. While expressed in endothelial and epithelial cells, occludin (62 - 82 KDa protein) is the functional component and the first transmembrane TJ protein identified. Furthermore, isoforms of occludins are the results of alternative splicing and the use of alternative promoters (Zhu et al. 2015). The carboxyl terminal of occludin forms a coiled structure of dimers that associate with different ZO-proteins within the junction complex (Wittchen et al. 1999; Fanning et al. 1998; Itoh et al. 1999).

Tricellulin is the first TJ protein abundant at tricellular TJ (TTJ) (Ikenouchi et al. 2005). In contrast to occludin, tricellulin contains an extended N-terminus and C-terminus of 194 and 195 amino acids, respectively, contributing to the stabilization of the TJs. During the epithelial-mesenchymal transition (EMT), the expression of tricellulin decreases. During cell apoptosis, tricellulin is cleaved at Asp-441 and Asp-487 residues, resulting in the dissociation of TTJs (Janke et al. 2019). In addition to cancer, a mutation in the TRIC gene has been associated with deafness (Mariano et al. 2011).

c. Junctional adhesion molecules
Junctional adhesion molecules (JAMs) is a special family of 32 KDa type I transmembrane proteins with immuno-globin-like ectodomains. JAMs are adhesion molecules that are present at the TJs of polarized cells and on the cell surfaces of leukocytes, immune, and myocardial cells (Martìn-Padura et al. 1998; Liu et al. 2000). JAM-A contains two extracellular immunoglobulin-like loops, a single transmembrane, and a cytoplasmic domain ending in PDZ binding motif. An atypical protein kinase C (aPKC) phosphorylates JAM-A at Ser-285, which is important for the normal function of TJ (Iden et al. 2012).

2. Scaffolding proteins of tight junctions
   a. Zonular occludens (ZO-1, 2 and 3): Among the members, ZO-1 (220 kDa), a peripheral membrane molecule, was the first identified and the best-studied TJ protein. ZO-1, 2, and 3 are multi-domain proteins, each with three N-terminal PDZ (PSD95, DlgA, ZO1 homology) domains, a central SRC homology 3 (SH3) domain, and a region with homology to guanylate kinase (GUK) (Jesaitis and Goodenough 1994; Haskins et al. 1998; Willott et al. 1993). All three ZO proteins bind cytoskeleton actin and many actin-binding proteins. The carboxy terminus consisting of alternatively spliced domains is essential for interacting with the F-actin and would confer tissue-specific functions (Wittchen et al. 1999; Fanning et al. 1998, 2002).

b. Cingulin: Cingulin is a 140 kDa protein found in the TJs of both endothelial and epithelial cells. Cingulin is a coiled domain-containing protein that directly interacts with ZO-1. The phosphorylation of cingulin helps the junctional association of the microtubules, GTPase regulatory protein recruitment, and junction formation (Fanning and Anderson 2009; Aijaz et al. 2005).
c. Afadin: Afadin protein, present in TJ and adherens junctions, is essential for early polarization of the apical junctional complex (Guillemot et al. 2014). Disturbances resulting in unusual positioning and coordination between TJs and adherens affect function more importantly during development and remodeling. Similar to ZO, afadin also contains a PDZ domain and binds to both transmembrane proteins and actin filaments. Recent studies have shown that afadin, nectin, and ZO1 play an important role in the proper apical positioning of the TJs (Huang et al. 2012).
d. Membrane-associated guanylate kinase inverted:
Membrane-associated guanylate kinase inverted MAGI-1 and -3 are a class of proteins containing the PDZ domain (Severson et al. 2009). These scaffold proteins are localized at the epithelial TJs, with MAGI-3 present at other cellular locations (Ide et al. 1999). While MAGI-1 interacts with JAM-4, PTEN homolog, PDZ-GEF1, and actin-binding proteins (α-actinin 4 and synaptopodin), MAGI-3 functions as a scaffold in the WNT signaling pathway (Sakurai et al. 2006). Despite all this, MAGI exact and actual role is yet unraveled (Adamsky et al. 2003; Yao et al. 2004; Hirabayashi et al. 2003; Wu et al. 2000a).
e. Multiple PDZ proteins: Multiple PDZ proteins (MPDZ) have 13 PDZ domains and are present at TJs (Hamazaki et al. 2002). It interacts with various TJ proteins, including claudins, JAMs, etc. (Hamazaki et al. 2002; Jeansonne et al. 2003). This depicts a clear role of MPDZ proteins as a synaptic scaffold at tight in epithelial cells. One unusual association of MPDZ is that it acts as a scaffold somatostatin receptor at some tight epithelial junctions that can affect the permeability and biogenesis of these biological barriers (Liew et al. 2009).
f. Barmotin/7H6 is a 155 kDa protein within the TJ of hepatocytes, epithelial cells, and endothelial cells. It is thought to regulate paracellular permeability and barrier function.

1. Functions of tight junctions:
   a. Physiological roles of tight junctions
   Tight junctions are essential for cellular physiological processes, such as bidirectional cell signalling, which is involved in cell growth, proliferation, differentiation, migration, and survival. The functionality and the dual nature of the TJs are highly affected by the molecular composition or architecture. This molecular composition varies significantly between different epithelia. Tight junctions may act as the barrier or channel for the particular solute. Thus, differential expression of TJ proteins is under tight control by different transcriptional factors, including ZONAB. ZONAB is a transcriptional factor that regulates epithelial cell proliferation and shuttle between TJs and the nucleus

(Balda et al. 2003; Lima et al. 2010). Grainy head-like proteins GRHL-1 and -2 or the nuclear receptors control the expression of different junction proteins (Boivin and Schmidt-Ott 2017; Werth et al. 2010). These transcriptional factors control a significant number of the proteins that form intricate apical junctions. More often, a single factor can regulate the expression of multiple genes of the apical junction complex. GRHL2 regulates epithelial differentiation, as it controls the expression of components of both apical junctions, including claudins.

b. The barriers between the blood and the brain, the spinal cord, and the retina
Some vital body parts are protected from direct contact with the blood supply. They are protected by TJs as blood barriers from free influx and outflux of intra- and extracellular substances. Thus, they provide an optimal medium and conditions for normal function and protect it from fluctuations in ionic composition. The blood–brain barrier is a selective semipermeable membrane that separates blood in the brain and spinal cord from extracellular substances. It prevents the free passage of cells and blood substances. Normal transport of substances between brain cells and blood and between brain cells occurs through specific ion transporters and channels.

The blood–retinal barrier prevents retinal cells from transporting immune privilege to the ocular microenvironment. The blood–retina barrier (BRB) is made up of an outer barrier formed at the layer of retinal pigment epithelial (RPE) cells and an inner retinal barrier developed by TJs present in the retina's capillaries. The TJs positioned between retinal pigment epithelial (RPE) cells and Bruch membrane cells regulate the selective transport of substances from the blood to the tissues in order to maintain the homeostasis of the brain and spinal cord.

c. Tight junctions and MAP kinase and Ras signaling pathway
Gene expression, mitosis, differentiation, and apoptosis are regulated by MAP kinases, Ser/Thr protein kinases that react to external stimuli like stress and growth hormones. Mammals have four different types of MAPKs: the first is triggered by extracellular signal-regulated kinases (ERKs) that control cell division and proliferation. Occludin and ERK1 have been identified to interact in epithelial cells. The c-Jun N-terminal kinases (JNKs) and p38 isoforms, which are involved in cell differentiation and death, are the other two categories of MAPKs that are activated by stress stimuli. Both growth factors and stress-related stimuli have been identified to activate a fourth group, ERK5, which is involved in cell proliferation (González-Mariscal et al. 2008). TJ proteins have been found to be associated with different effector molecules of the Ras signaling pathway mediating cell proliferation. Raf kinases are the Ras effectors that lead to sequential activation of the kinase/mitogen-activated protein kinase (ERK/ MAPK) pathway and the TF Elk1. This sequential activation of kinases is sufficient for transforming epithelial cells (Pa4–Raf1) (Li and Mrsny 2000; Montesano et al. 1999). TJs seem to be connected with Raf-1 signaling as Raf-1 overexpression in epithelial cells represses the expression of occludin and induces an oncogenic phenotype by

downregulating the expression of occluding (Li and Mrsny 2000; Wang et al. 2005). It is observed that overexpression of occludins is enough to reverse the Raf-1 mediated transformation of a salivary-gland epithelial cell line Pa4 cells (Pa4-Raf1) (Li and Mrsny 2000; Wang et al. 2007). MAP signaling pathway downregulates several TJ proteins' expression and modulates paracellular transport. ERK1/2, a downstream molecule of Ras signaling activation by mutant Ras and HGF, increases the trans-epithelial permeability of mannitol and diminishes claudin-2 expression in the leaky strain of MDCK cells, respectively (Lipschutz et al. 2005). Thiols have also been observed to facilitate TJ sealing through the activation and overexpression of the ERK1/2, JNK, and p38 pathways (González-Mariscal et al. 2008). Moreover, glucocorticoid therapy raises TER by bringing Ras to areas of cell–cell interaction and activating ERK1/2 (Woo et al. 1999).

d. Role in gene expression and proliferation

TJs also play an essential role in regulating gene expression, cell growth, and proliferation. Much evidence about gene expression and cell proliferation comes from the TJ-associated adaptor proteins. One of the adaptor proteins, TJ plaque protein ZO-1, negatively controls cell development and is a member of the same protein family as the Drosophila Disc-large tumor suppressor (D1Ga) (Willott et al. 1993; Tsukita et al. 1993). Other proteins, such as ZO-2, ZO-3, and several other proteins along with ZO-1, are present at the TJs. It has been demonstrated that the TJ-associated ZO-1 and ZO-2 adaptor proteins control transcription factors and epithelial proliferation. The discovery that mutations in ZO-1's Drosophila homolog, Tamou/polychaetoid, induce abnormalities in dorsal closure, epithelial migration, and cell-fate determination in sensory organs suggests that ZO-1 may play a role in the regulation of cell differentiation (Takahisa et al. 1996; Chen et al. 1996). The evidence for the function of TJ proteins in regulating cell proliferation and differentiation came with the experiments in which TJs were manipulated either by the expression of truncated ZO-1 or the addition of inhibitory peptides against the extracellular domain of occluding. Both treatments resulted in partial differentiation (Lima et al. 2010; Ryeom et al. 2000; Vietor et al. 2001). In addition, other TJ adaptor proteins like MAGI-2 and MAGI-3 interact with tumor suppressor phosphatase and tensin homolog (PTEN) and inhibit signaling pathways associated with cell proliferation (Wu et al. 2000a,b). TJ-associated transcription factor ZONAB interacts with the Src-homology-3 (SH3) domain of zonula occludens-1 (ZO-1) and regulates the expression of proto-oncogene ErbB2 to control epithelial cell proliferation and cell density (Balda et al. 2003; Balda and Matter 2000). In addition, the complex formation of ZONAB with cyclin-dependent kinase 4 (CDK4) has been shown to promote its nuclear translocation and cell cycle progression (Balda et al. 2003). Therefore ZO-1, ZONAB, and CDK4 form a part of cell-density-dependent signaling that regulates gene expression and epithelial cell proliferation. In addition to gene regulations, components of TJ such as symplekin regulate polyadenylation machinery (Takagaki and Manley 2000; Hofmann et al. 2002), Y-box factor ZONAB binds DNA and RNA), and ZO-2 interacting protein SAF-B.

e. Regulation of permeability
TJs regulate the paracellular permeability of molecules by involving two distinct pathways. One is "Pore" pathway, which allows small charged and uncharged molecules, ions, and nutrients of ~4 Å in size to pass through the TJ along with water. The pore pathway is affected by claudins' expression, composition, and paracellular flux magnitude. The second pathway is called the "leaky" pathway that allows the movement of particles more than 4Å to pass through the TJs without the selectivity of charge. This pathway is controlled by cytoskeleton factors that affect cell homeostasis (Anderson and Van Itallie 2009). Many studies have reported that infection with *Helicobacter pylori* disrupts TJs and decreases trans-epithelial resistance (TER), increases permeability, and injures gastric epithelial cells among both these pathways.

f. Tight junctions and Helicobacter infection
The International Agency for Research on Cancer (IARC) and the World Health Organization (WHO) classify Helicobacter pylori as a type 1 carcinogen (IARC) (Author 1994). It is more frequently linked to stomach cancer, the third most frequent cause of cancer-related death worldwide and is ranked as the fifth most prevalent cancer overall (Ferlay et al. 2015; Colquhoun et al. 2015). *H. pylori* leads to intestinal-type gastric adenocarcinoma and initiates a well-defined pathological process called "cornea cascade," characterized by chronic superficial gastritis followed by atrophic gastritis and intestinal metaplasia which progresses to dysplasia and adenocarcinoma (Correa 1992; Correa and Houghton 2007). Eradication of *H. pylori* does not reduce the metachronous gastric carcinoma (Choi et al. 2014) and suggests irreversible genetic, epigenetic, or other permanent changes. TJ function is regulated by phosphorylated myosin regulatory light chain (MLC). Interestingly, *H. pylori* were shown to activate MLC kinase, promoting MLC phosphorylation with progressive loss of TER. The underlying mechanism revealed the production of ammonia from urea by *H. pylori* enzyme urease, which results in the activation of MLC kinase (Yuhan et al. 1997) and modulates the expression of claudin-18 (Yuhan et al. 1997). MLC activation and subsequent activation of claudins increases intestinal paracellular permeability and often accompanies precedence inflammatory bowel disease, graft versus host disease, and infectious enterocolitis (Yuhan et al. 1997; Clayburgh et al. 2005; Scott et al. 2002).

7.4 Disorders Related to TJs

Many tissues and organ disorders are associated with the deformity or dysregulation of TJs. Some diseases result from genetic origin and involve different mutations or polymorphisms in the TJ-associated proteins resulting in the dysregulation of different TJ-associated signaling pathways, for example, in cystic fibrosis. In addition, pathogenic microbes directly interact with a particular protein of the TJ complex and result in the activation of the signaling pathways.

TJ, as an adhesive, prevents cell dissociation in epithelial cells (Hollande et al. 2001). Additionally, it has been discovered that establishing and developing cancer metastasis involves the interaction and penetration of dissociated cancer cells through the vascular endothelium. The dysregulation of TJ proteins has been strongly implicated in the progression and metastatic potential of many tumors, including breast, liver, and ovarian cancer. For example, claudins play a crucial role in TJ cancer signaling. Claudin 1 phosphorylation by MAP kinase (Fujibe et al. 2004) or by protein kinase C and the claudin 5 phosphorylation by cyclic AMP promote the barrier functions of TJs (Ishizaki et al. 2003) and increase paracellular permeability (Yamauchi et al. 2004). However, abnormal phosphorylation of claudins affects their aggregation and structural stability and impairs their epithelial barrier function (Sjö et al. 2010). The metastasis involves many steps, including invasion of the tumor cells into the surrounding stroma, survival through the body circulation, angiogenesis, and development of the tumor that requires delivery of nutrients and removal of waste products from the tumor site. Cells that are resistant to anoikis experience anchorage-independent growth and EMT, which are essential for the progression and spread of cancer (Sjö et al. 2010; Philip et al. 2015). It has been observed that Akt, a signaling protein plays a central role in anoikis resistance by decreasing the pro-apoptotic proteins and increasing the anti-apoptotic proteins (Jeong et al. 2008). Overexpression of claudin-2 in tissue samples from colorectal cancer (CRC) patients was correlated to cancer progression. In CRC cells, it has been discovered that claudin-2 overexpression promoted cell proliferation and anchorage-independent tumor growth in a way that is dependent on the EGF receptor regulation (Dhawan et al. 2011, 2005). Similarly, increased claudin 3 expression promoted ERK1/2 and PI3-Akt signaling pathways and increased cell migration and anchorage-independent behavior of human CRC. The Notch pathway is activated by the overexpression of claudin-1, which also causes mucosal inflammation and increases the propensity for metastasis by changing the expression of E-cadherin and Wnt/β catenin (Dhawan et al. 2005; Pope et al. 2014). In contrast, decreased expression of claudins like claudin-7 has been observed in invasive breast and head and neck cancers.

In addition to cancer, claudins' mutations are associated with various diseases. For example, mutation of claudin -16 in the ascending limb of the loop of Henle is associated with increased urinary loss of Mg++ and Ca++ and reduced plasma Mg+

+ levels, leading to weakness and seizures (Anderson and Van Itallie 2009). Similarly, altered claudin-19 expression in the retina leads to vision loss (Konrad et al. 2006), and claudin-1 and -14 mutant causes hearing loss and neonatal ichthyosis-sclerosing cholangitis syndrome (NISCH), respectively (Feldmeyer et al. 2006). It is clear from multiple pieces of literature that the TJ and TJ-associated proteins are essential in regulating normal cell physiology. TJs-associated proteins have been linked to various signaling and transcriptional pathways regulating cell proliferation and gene expression differentiation. It is also clear that disruption of TJ and its associated proteins led to multiple diseases like cancer and inflammation. The future direction of this study will be the complete understanding of the TJs supramolecular complex by revealing the architecture of TJs. The TJ-associated proteins could potentially be used as new therapeutic targets and novel biomarkers in cancers.

7.5 Conclusion

While the structure and function of TJs are continually evolving, a vision of TJs as supramolecular cell–cell adhesion complexes that assemble claudins and other membrane proteins associated is already well understood. To answer meaningful questions regarding how this junctional complex is organized and physiologically regulated, it requires structural data of the TJs integral membrane proteins and other associated proteins. A lipid and protein molecular models of TJs seek to explain the structural and functional properties. All cells use cell adhesion molecules as a universal method for facilitating communication and interaction. TJ proteins act as barriers that impede the diffusion of various substances and act as a boundary wall between apical and basolateral plasma membrane domains. The TJs-associated transmembrane proteins bind and recruit various signaling molecules involved in multiple cellular processes, including cell proliferation, migration, gene expression, and coordinates junction assembly. Integral membrane proteins and scaffolding proteins of TJs are not only static proteins but also accomplish the development of apical-basolateral polarity and control of paracellular trafficking of ions and solutes.

Moreover, the overexpression of TJs-associated proteins led to different inflammatory diseases and cancers. TJs are desired diagnostic and prognostic markers because they are essential for tissue compartmentalization and cellular homeostasis; any abnormality in them signifies a variety of diseases. In conditions of illness, such as barrier disruption during infection and inflammation to loss of cell polarity and altered migratory behavior during cancer, there is a developing understanding of the interaction of the numerous TJ proteins and their dysregulation. Overall, TJs architecture are important in multiple cellular functions, and their dysregulation leads to several pathophysiological conditions (Table 7.1).

Table 7.1 Disorders coupled with tight junction proteins

S. No.	Junctional proteins	Associated genetic disease	Mechanism	References
Transmembrane proteins				
1	Claudin 1	Neonatal ichthyosis, neonatal sclerosing cholangitis	Absence of claudin-1 causes liver damage and increased paracellular permeability	(Grosse et al. 2012; Hadj-Rabia et al. 2004)
2	Claudin 5	Velo-cardio-facial syndrome, schizophrenia	Physical symptoms include malformations of the palate and the heart and are brought on by a microdeletion in the long arm of chromosome 22	(Greene et al. 2018; Morita et al. 1999)
3	Claudin 14, tricellulin	Non-syndromic deafness	Mutations in the claudin-14 protein result in the mass apoptosis of hair cells, which results in autosomal recessive non-syndromic hearing loss	(Ben-Yosef et al. 2003)
4	Claudin 16 (paracellin 1)	Familial hypomagnesemia, hypercalciuria, nephrocalcinosis	A tubular illness that is genetically transmitted and is brought on by mutations in the claudin-16 or claudin-19 genes	(Vianna et al. 2019)
5	Claudin 19	Familial hypomagnesaemia, hypercalciuria, nephrocalcinosis, visual impairment	Retinal pigment epithelium and retinal differentiation are impacted at distinct stages by mutated claudin-19	(Vianna et al. 2019; Wang et al. 2019)
Adaptor proteins				
1	ZO2	Familial hypercholanemia	Mutations in the gene encoding TJP2, also known as ZO-2, are the cause of familial hypercholanemia (FHC), which is characterized by increased blood bile acid concentrations, pruritus, and fat malabsorption	(Carlton et al. 2003)

(continued)

Table 7.1 (continued)

S. No.	Junctional proteins	Associated genetic disease	Mechanism	References
Signaling proteins				
1	p114RHOGEF	Clarkson disease, retinitis pigmentosa, non-idiopathic pulmonary arterial hypertension susceptibility	Inactivating mutations in p114RhoGEF	(Arno et al. 2017; Beal et al. 2021)
2	ZONAB	Activated in cystic fibrosis owing to downregulation of ZO1, a binding partner of CFTR	By altering the ZO-1-ZONAB pathway, CFTR controls tight junction construction and epithelial cell differentiation	(Ruan et al. 2014)
3	WNK4 (a kinase that phosphorylates claudins)	Pseudohypoaldosteronism type I and II	PHA II is caused by mutations in WNK1 and WNK4, and PHA II mutations in WNK4 reverse the negative regulation of the NaCl cotransporter (NCC) that WNK4 has	(Furgeson and Linas 2010)
4	Junctional protein targeted by pathogen	Pathogenic virus or bacterium (implications)		
Transmembrane proteins				
1	Claudin 1, claudin 6, claudin 9, occluding	Hepatitis C virus (infection)	Claudin-1, -2, -4, -6, and -9 function as cofactors for HCV entrance	(Meertens et al. 2008)
2	JAMA	Reoviruses (infection)	JAM-A facilitates the entry of reovirus	(Campbell et al. 2005)
3	CAR Coxsackie viruses and adenoviruses receptor	Cancers, myocarditis, and dilated cardiomyopathy	DCM and myocarditis are caused by high CAR expression, while bladder cell cancer is caused by low CAR expression	(Matsumoto et al. 2005; Kaur et al. 2012)
4	Claudin 3, claudin 4	Clostridium perfringens (junction dissociation)	A common enterotoxin (CpE) produced by Clostridium perfringens targets claudins which causes gastrointestinal problems	(Ogbu et al. 2022)
5	Occludin	Vibrio cholerae (junction dissociation)	Through enterotoxin, Vibrio cholera has been demonstrated to disrupt occludin TJs	(Groschwitz and Hogan 2009)

(continued)

Table 7.1 (continued)

S. No.	Junctional proteins	Associated genetic disease	Mechanism	References
Adaptor proteins				
1	ZO1, ZO2	Tick-borne encephalitis and dengue viruses	The flaviviruses dengue virus (DENV) and tick-borne encephalitis virus (TBEV) are capable of inflicting fatal hemorrhagic fever and encephalitis, respectively, through cell proliferation, junction permeability, and the interferon pathways	(Ellencrona et al. 2009)
2	ZO2, MUPP1, PATJ, MAGI1	Adenoviruses (arrhythmias in infected hearts, oncogenic variants, transformation)	Adenovirus type 5 (Ad5) infection causes reduced Cx43 gap junction protein levels which leads into loss of gap junction function. The human adenovirus E4-ORF1protein subverts membrane-associated cellular PDZ protein Discs Large 1 (The septate junction tumour suppressor protein of Drosophila discs-large 1 is homologous to the tight junction protein ZO-1) to mediate membrane recruitment and dysregulation of phosphatidylinositol 3-kinase	(Calhoun et al. 2020; Kong et al. 2014)
3	MAGI1–3, PATJ, MUPP1, PAR3	Papillomaviruses (papilloma formation)		
4	MAGI1–3	Influenza A virus (junction dissociation)		
5	PALS1	Junction development is delayed as a result of severe acute respiratory syndrome viral targeting		

(continued)

Table 7.1 (continued)

S. No.	Junctional proteins	Associated genetic disease	Mechanism	References
Signaling proteins				
1	GEFH1	Helicobacter pylori through protease-activated receptor-1 leads to displacement of structural proteins such as ZO1 and occludin		(Zihni et al. 2016)

Acknowledgment This study was supported by Ramalingaswami Re-entry Fellowship (Grant number: D.O. NO.BT/HRD/35/02/2006) from the Department of Biotechnology, Govt. of India, New Delhi and Core Research Grant (CRG/2021/003805) from Science and Engineering Research Board (SERB), Govt. of India, New Delhi to Muzafar A. Macha and Promotion of University Research and Scientific Excellence (PURSE) Grant to the Islamic University of Science and Technology, Awantipora from Department of Biotechnology, Govt. of India. The authors would like to acknowledge Sidra Medicine Precision Program for funding and constant support to Ajaz A. Bhat and Ammira S. Al-Shabeeb Akil.

Declarations **Ethical Approval and Consent to participate**
Not applicable.

Consent for Publication Not applicable.

Availability of Supporting Data Not applicable.

Competing Interests The authors declare that they have no competing interests.

References

Adamsky K, Arnold K, Sabanay H, Peles E (2003) Junctional protein MAGI-3 interacts with receptor tyrosine phosphatase beta (RPTP beta) and tyrosine-phosphorylated proteins. J Cell Sci 116(Pt 7):1279–1289

Aijaz S, D'Atri F, Citi S, Balda MS, Matter K (2005) Binding of GEF-H1 to the tight junction-associated adaptor cingulin results in inhibition of Rho signaling and G1/S phase transition. Dev Cell 8(5):777–786

Anderson JM, Van Itallie CM (2009) Physiology and function of the tight junction. Cold Spring Harb Perspect Biol 1(2):a002584

Arno G, Carss KJ, Hull S, Zihni C, Robson AG, Fiorentino A, Hardcastle AJ, Holder GE, Cheetham ME, Plagnol V et al (2017) Biallelic mutation of ARHGEF18, involved in the determination of epithelial apicobasal polarity, causes adult-onset retinal degeneration. Am J Hum Genet 100(2): 334–342

Author (1994) Infection with Helicobacter pylori. IARC Monogr Eval Carcinog Risks Hum 61: 177–240

Balda MS, Matter K (2000) The tight junction protein ZO-1 and an interacting transcription factor regulate ErbB-2 expression. EMBO J 19(9):2024–2033

Balda MS, Garrett MD, Matter K (2003) The ZO-1-associated Y-box factor ZONAB regulates epithelial cell proliferation and cell density. J Cell Biol 160(3):423–432

Beal R, Alonso-Carriazo Fernandez A, Grammatopoulos DK, Matter K, Balda MS (2021) ARHGEF18/p114RhoGEF coordinates PKA/CREB signaling and actomyosin remodeling to promote trophoblast cell-cell fusion during placenta morphogenesis. Front Cell Dev Biol 9: 658006

Ben-Yosef T, Belyantseva IA, Saunders TL, Hughes ED, Kawamoto K, Van Itallie CM, Beyer LA, Halsey K, Gardner DJ, Wilcox ER et al (2003) Claudin 14 knockout mice, a model for autosomal recessive deafness DFNB29, are deaf due to cochlear hair cell degeneration. Hum Mol Genet 12(16):2049–2061

Boivin FJ, Schmidt-Ott KM (2017) Transcriptional mechanisms coordinating tight junction assembly during epithelial differentiation. Ann N Y Acad Sci 1397(1):80–99

Calhoun PJ, Phan AV, Taylor JD, James CC, Padget RL, Zeitz MJ, Smyth JW (2020) Adenovirus targets transcriptional and posttranslational mechanisms to limit gap junction function. FASEB J 34(7):9694–9712

Campbell JA, Schelling P, Wetzel JD, Johnson EM, Forrest JC, Wilson GA, Aurrand-Lions M, Imhof BA, Stehle T, Dermody TS (2005) Junctional adhesion molecule a serves as a receptor for prototype and field-isolate strains of mammalian reovirus. J Virol 79(13):7967–7978

Carlton VE, Harris BZ, Puffenberger EG, Batta AK, Knisely AS, Robinson DL, Strauss KA, Shneider BL, Lim WA, Salen G et al (2003) Complex inheritance of familial hypercholanemia with associated mutations in TJP2 and BAAT. Nat Genet 34(1):91–96

Chen CM, Freedman JA, Bettler DR Jr, Manning SD, Giep SN, Steiner J, Ellis HM (1996) Polychaetoid is required to restrict segregation of sensory organ precursors from proneural clusters in Drosophila. Mech Dev 57(2):215–227

Chen Y, Merzdorf C, Paul DL, Goodenough DA (1997) COOH terminus of occludin is required for tight junction barrier function in early Xenopus embryos. J Cell Biol 138(4):891–899

Chernomordik LV, Kozlov MM (2005) Membrane hemifusion: crossing a chasm in two leaps. Cell 123(3):375–382

Choi J, Kim SG, Yoon H, Im JP, Kim JS, Kim WH, Jung HC (2014) Eradication of Helicobacter pylori after endoscopic resection of gastric tumors does not reduce incidence of metachronous gastric carcinoma. Clin Gastroenterol Hepatol 12(5):793–800

Citi S, Sabanay H, Jakes R, Geiger B, Kendrick-Jones J (1988) Cingulin, a new peripheral component of tight junctions. Nature 333(6170):272–276

Clayburgh DR, Barrett TA, Tang Y, Meddings JB, Van Eldik LJ, Watterson DM, Clarke LL, Mrsny RJ, Turner JR (2005) Epithelial myosin light chain kinase-dependent barrier dysfunction mediates T cell activation-induced diarrhea in vivo. J Clin Invest 115(10):2702–2715

Colquhoun A, Arnold M, Ferlay J, Goodman KJ, Forman D, Soerjomataram I (2015) Global patterns of cardia and non-cardia gastric cancer incidence in 2012. Gut 64(12):1881–1888

Cording J, Berg J, Käding N, Bellmann C, Tscheik C, Westphal JK, Milatz S, Günzel D, Wolburg H, Piontek J et al (2013) In tight junctions, claudins regulate the interactions between occludin, tricellulin and marvelD3, which, inversely, modulate claudin oligomerization. J Cell Sci 126(Pt 2):554–564

Correa P (1992) Human gastric carcinogenesis: a multistep and multifactorial process–First American Cancer Society Award Lecture on Cancer Epidemiology and Prevention. Cancer Res 52(24):6735–6740

Correa P, Houghton J (2007) Carcinogenesis of Helicobacter pylori. Gastroenterology 133(2): 659–672

Dhawan P, Singh AB, Deane NG, No Y, Shiou SR, Schmidt C, Neff J, Washington MK, Beauchamp RD (2005) Claudin-1 regulates cellular transformation and metastatic behavior in colon cancer. J Clin Invest 115(7):1765–1776

Dhawan P, Ahmad R, Chaturvedi R, Smith JJ, Midha R, Mittal MK, Krishnan M, Chen X, Eschrich S, Yeatman TJ et al (2011) Claudin-2 expression increases tumorigenicity of colon cancer cells: role of epidermal growth factor receptor activation. Oncogene 30(29):3234–3247

Ellencrona K, Syed A, Johansson M (2009) Flavivirus NS5 associates with host-cell proteins zonula occludens-1 (ZO-1) and regulating synaptic membrane exocytosis-2 (RIMS2) via an internal PDZ binding mechanism. Biol Chem 390(4):319–323

Fanning AS, Anderson JM (2009) Zonula occludens-1 and -2 are cytosolic scaffolds that regulate the assembly of cellular junctions. Ann N Y Acad Sci 1165:113–120

Fanning AS, Jameson BJ, Jesaitis LA, Anderson JM (1998) The tight junction protein ZO-1 establishes a link between the transmembrane protein occludin and the actin cytoskeleton. J Biol Chem 273(45):29745–29753

Fanning AS, Ma TY, Anderson JM (2002) Isolation and functional characterization of the actin binding region in the tight junction protein ZO-1. FASEB J 16(13):1835–1837

Farquhar MG, Palade GE (1963) Junctional complexes in various epithelia. J Cell Biol 17(2): 375–412

Feldmeyer L, Huber M, Fellmann F, Beckmann JS, Frenk E, Hohl D (2006) Confirmation of the origin of NISCH syndrome. Hum Mutat 27(5):408–410

Ferlay J, Soerjomataram I, Dikshit R, Eser S, Mathers C, Rebelo M, Parkin DM, Forman D, Bray F (2015) Cancer incidence and mortality worldwide: sources, methods and major patterns in GLOBOCAN 2012. Int J Cancer 136(5):E359–E386

Fernández Miyakawa ME, Pistone Creydt V, Uzal FA, McClane BA, Ibarra C (2005) Clostridium perfringens enterotoxin damages the human intestine in vitro. Infect Immun 73(12):8407–8410

Fujibe M, Chiba H, Kojima T, Soma T, Wada T, Yamashita T, Sawada N (2004) Thr203 of claudin-1, a putative phosphorylation site for MAP kinase, is required to promote the barrier function of tight junctions. Exp Cell Res 295(1):36–47

Fujimoto K (1995) Freeze-fracture replica electron microscopy combined with SDS digestion for cytochemical labeling of integral membrane proteins. Application to the immunogold labeling of intercellular junctional complexes. J Cell Sci 108(Pt 11):3443–3449

Furgeson SB, Linas S (2010) Mechanisms of type I and type II pseudohypoaldosteronism. J Am Soc Nephrol 21(11):1842–1845

Furuse M, Hirase T, Itoh M, Nagafuchi A, Yonemura S, Tsukita S, Tsukita S (1993) Occludin: a novel integral membrane protein localizing at tight junctions. J Cell Biol 123(6):1777–1788

Furuse M, Sasaki H, Fujimoto K, Tsukita S (1998) A single gene product, claudin-1 or -2, reconstitutes tight junction strands and recruits occludin in fibroblasts. J Cell Biol 143(2): 391–401

Furuse M, Sasaki H, Tsukita S (1999) Manner of interaction of heterogeneous claudin species within and between tight junction strands. J Cell Biol 147(4):891–903

Gonçalves A, Ambrósio AF, Fernandes R (2013) Regulation of claudins in blood-tissue barriers under physiological and pathological states. Tissue Barriers 1(3):e24782

González-Mariscal L, Tapia R, Chamorro D (2008) Crosstalk of tight junction components with signaling pathways. Biochim Biophys Acta 1778(3):729–756

Greene C, Kealy J, Humphries MM, Gong Y, Hou J, Hudson N, Cassidy LM, Martiniano R, Shashi V, Hooper SR et al (2018) Dose-dependent expression of claudin-5 is a modifying factor in schizophrenia. Mol Psychiatry 23(11):2156–2166

Groschwitz KR, Hogan SP (2009) Intestinal barrier function: molecular regulation and disease pathogenesis. J Allergy Clin Immunol 124(1):3–20; quiz 21-22.

Grosse B, Cassio D, Yousef N, Bernardo C, Jacquemin E, Gonzales E (2012) Claudin-1 involved in neonatal ichthyosis sclerosing cholangitis syndrome regulates hepatic paracellular permeability. Hepatology 55(4):1249–1259

Guillemot L, Guerrera D, Spadaro D, Tapia R, Jond L, Citi S (2014) MgcRacGAP interacts with cingulin and paracingulin to regulate Rac1 activation and development of the tight junction barrier during epithelial junction assembly. Mol Biol Cell 25(13):1995–2005

Günzel D, Fromm M (2012) Claudins and other tight junction proteins. Compr Physiol 2(3): 1819–1852

Günzel D, Yu AS (2013) Claudins and the modulation of tight junction permeability. Physiol Rev 93(2):525–569

Hadj-Rabia S, Baala L, Vabres P, Hamel-Teillac D, Jacquemin E, Fabre M, Lyonnet S, De Prost Y, Munnich A, Hadchouel M et al (2004) Claudin-1 gene mutations in neonatal sclerosing cholangitis associated with ichthyosis: a tight junction disease. Gastroenterology 127(5): 1386–1390

Hamazaki Y, Itoh M, Sasaki H, Furuse M, Tsukita S (2002) Multi-PDZ domain protein 1 (MUPP1) is concentrated at tight junctions through its possible interaction with claudin-1 and junctional adhesion molecule. J Biol Chem 277(1):455–461

Haseloff RF, Dithmer S, Winkler L, Wolburg H, Blasig IE (2015) Transmembrane proteins of the tight junctions at the blood-brain barrier: structural and functional aspects. Semin Cell Dev Biol 38:16–25

Haskins J, Gu L, Wittchen ES, Hibbard J, Stevenson BR (1998) ZO-3, a novel member of the MAGUK protein family found at the tight junction, interacts with ZO-1 and occludin. J Cell Biol 141(1):199–208

Hirabayashi S, Tajima M, Yao I, Nishimura W, Mori H, Hata Y (2003) JAM4, a junctional cell adhesion molecule interacting with a tight junction protein, MAGI-1. Mol Cell Biol 23(12): 4267–4282

Hofmann I, Schnölzer M, Kaufmann I, Franke WW (2002) Symplekin, a constitutive protein of karyo- and cytoplasmic particles involved in mRNA biogenesis in Xenopus laevis oocytes. Mol Biol Cell 13(5):1665–1676

Hollande F, Blanc EM, Bali JP, Whitehead RH, Pelegrin A, Baldwin GS, Choquet A (2001) HGF regulates tight junctions in new nontumorigenic gastric epithelial cell line. Am J Physiol Gastrointest Liver Physiol 280(5):G910–G921

Huang RY, Guilford P, Thiery JP (2012) Early events in cell adhesion and polarity during epithelial-mesenchymal transition. J Cell Sci 125(Pt 19):4417–4422

Ide N, Hata Y, Nishioka H, Hirao K, Yao I, Deguchi M, Mizoguchi A, Nishimori H, Tokino T, Nakamura Y et al (1999) Localization of membrane-associated guanylate kinase (MAGI)-1/ BAI-associated protein (BAP) 1 at tight junctions of epithelial cells. Oncogene 18(54): 7810–7815

Iden S, Misselwitz S, Peddibhotla SS, Tuncay H, Rehder D, Gerke V, Robenek H, Suzuki A, Ebnet K (2012) aPKC phosphorylates JAM-A at Ser285 to promote cell contact maturation and tight junction formation. J Cell Biol 196(5):623–639

Ikenouchi J, Furuse M, Furuse K, Sasaki H, Tsukita S, Tsukita S (2005) Tricellulin constitutes a novel barrier at tricellular contacts of epithelial cells. J Cell Biol 171(6):939–945

Ishizaki T, Chiba H, Kojima T, Fujibe M, Soma T, Miyajima H, Nagasawa K, Wada I, Sawada N (2003) Cyclic AMP induces phosphorylation of claudin-5 immunoprecipitates and expression of claudin-5 gene in blood-brain-barrier endothelial cells via protein kinase A-dependent and -independent pathways. Exp Cell Res 290(2):275–288

Itoh M, Morita K, Tsukita S (1999) Characterization of ZO-2 as a MAGUK family member associated with tight as well as adherens junctions with a binding affinity to occludin and alpha catenin. J Biol Chem 274(9):5981–5986

Janke S, Mittag S, Reiche J, Huber O (2019) Apoptotic fragmentation of tricellulin. Int J Mol Sci 20:19

Jeansonne B, Lu Q, Goodenough DA, Chen YH (2003) Claudin-8 interacts with multi-PDZ domain protein 1 (MUPP1) and reduces paracellular conductance in epithelial cells. Cell Mol Biol 49(1): 13–21

Jeong SJ, Dasgupta A, Jung KJ, Um JH, Burke A, Park HU, Brady JN (2008) PI3K/AKT inhibition induces caspase-dependent apoptosis in HTLV-1-transformed cells. Virology 370(2):264–272

Jesaitis LA, Goodenough DA (1994) Molecular characterization and tissue distribution of ZO-2, a tight junction protein homologous to ZO-1 and the Drosophila discs-large tumor suppressor protein. J Cell Biol 124(6):949–961

Kan FW (1993) Cytochemical evidence for the presence of phospholipids in epithelial tight junction strands. J Histochem Cytochem 41(5):649–656

Kaur T, Mishra B, Saikia UN, Sharma M, Bahl A, Ratho RK (2012) Expression of coxsackievirus and adenovirus receptor and its cellular localization in myocardial tissues of dilated cardiomyopathy. Exp Clin Cardiol 17(4):183–186

Kong K, Kumar M, Taruishi M, Javier RT (2014) The human adenovirus E4-ORF1 protein subverts discs large 1 to mediate membrane recruitment and dysregulation of phosphatidylinositol 3-kinase. PLoS Pathog 10(5):e1004102

Konrad M, Schaller A, Seelow D, Pandey AV, Waldegger S, Lesslauer A, Vitzthum H, Suzuki Y, Luk JM, Becker C et al (2006) Mutations in the tight-junction gene claudin 19 (CLDN19) are associated with renal magnesium wasting, renal failure, and severe ocular involvement. Am J Hum Genet 79(5):949–957

Krause G, Winkler L, Mueller SL, Haseloff RF, Piontek J, Blasig IE (2008) Structure and function of claudins. Biochim Biophys Acta 1778(3):631–645

Lambert D, O'Neill CA, Padfield PJ (2005) Depletion of Caco-2 cell cholesterol disrupts barrier function by altering the detergent solubility and distribution of specific tight-junction proteins. Biochem J 387(Pt 2):553–560

Li D, Mrsny RJ (2000) Oncogenic Raf-1 disrupts epithelial tight junctions via downregulation of occludin. J Cell Biol 148(4):791–800

Liew CW, Vockel M, Glassmeier G, Brandner JM, Fernandez-Ballester GJ, Schwarz JR, Schulz S, Buck F, Serrano L, Richter D et al (2009) Interaction of the human somatostatin receptor 3 with the multiple PDZ domain protein MUPP1 enables somatostatin to control permeability of epithelial tight junctions. FEBS Lett 583(1):49–54

Lima WR, Parreira KS, Devuyst O, Caplanusi A, N'Kuli F, Marien B, Van Der Smissen P, Alves PM, Verroust P, Christensen EI et al (2010) ZONAB promotes proliferation and represses differentiation of proximal tubule epithelial cells. J Am Soc Nephrol 21(3):478–488

Lipschutz JH, Li S, Arisco A, Balkovetz DF (2005) Extracellular signal-regulated kinases 1/2 control claudin-2 expression in Madin-Darby canine kidney strain I and II cells. J Biol Chem 280(5):3780–3788

Liu Y, Nusrat A, Schnell FJ, Reaves TA, Walsh S, Pochet M, Parkos CA (2000) Human junction adhesion molecule regulates tight junction resealing in epithelia. J Cell Sci 113(Pt 13): 2363–2374

Lynch RD, Francis SA, McCarthy KM, Casas E, Thiele C, Schneeberger EE (2007) Cholesterol depletion alters detergent-specific solubility profiles of selected tight junction proteins and the phosphorylation of occludin. Exp Cell Res 313(12):2597–2610

Mariano C, Sasaki H, Brites D, Brito MA (2011) A look at tricellulin and its role in tight junction formation and maintenance. Eur J Cell Biol 90(10):787–796

Martìn-Padura I, Lostaglio S, Schneemann M, Williams L, Romano M, Fruscella P, Panzeri C, Stoppacciaro A, Ruco L, Villa A et al (1998) Junctional adhesion molecule, a novel member of the immunoglobulin superfamily that distributes at intercellular junctions and modulates monocyte transmigration. J Cell Biol 142(1):117–127

Matsumoto K, Shariat SF, Ayala GE, Rauen KA, Lerner SP (2005) Loss of coxsackie and adenovirus receptor expression is associated with features of aggressive bladder cancer. Urology 66(2):441–446

Meertens L, Bertaux C, Cukierman L, Cormier E, Lavillette D, Cosset FL, Dragic T (2008) The tight junction proteins claudin-1, -6, and -9 are entry cofactors for hepatitis C virus. J Virol 82(7):3555–3560

Mineta K, Yamamoto Y, Yamazaki Y, Tanaka H, Tada Y, Saito K, Tamura A, Igarashi M, Endo T, Takeuchi K et al (2011) Predicted expansion of the claudin multigene family. FEBS Lett 585(4): 606–612

Montesano R, Soriano JV, Hosseini G, Pepper MS, Schramek H (1999) Constitutively active mitogen-activated protein kinase kinase MEK1 disrupts morphogenesis and induces an invasive phenotype in Madin-Darby canine kidney epithelial cells. Cell Growth Differ 10(5):317–332

Morita K, Sasaki H, Furuse M, Tsukita S (1999) Endothelial claudin: claudin-5/TMVCF constitutes tight junction strands in endothelial cells. J Cell Biol 147(1):185–194

Nusrat A, Parkos CA, Verkade P, Foley CS, Liang TW, Innis-Whitehouse W, Eastburn KK, Madara JL (2000) Tight junctions are membrane microdomains. J Cell Sci 113(Pt 10): 1771–1781

Ogbu CP, Roy S, Vecchio AJ (2022) Disruption of claudin-made tight junction barriers by clostridium perfringens enterotoxin: insights from structural biology. Cell 11:5

Philip R, Heiler S, Mu W, Büchler MW, Zöller M, Thuma F (2015) Claudin-7 promotes the epithelial-mesenchymal transition in human colorectal cancer. Oncotarget 6(4):2046–2063

Pinto da Silva P, Kachar B (1982) On tight-junction structure. Cell 28(3):441–450

Pope JL, Ahmad R, Bhat AA, Washington MK, Singh AB, Dhawan P (2014) Claudin-1 overexpression in intestinal epithelial cells enhances susceptibility to adenamatous polyposis coli-mediated colon tumorigenesis. Mol Cancer 13:167

Ruan YC, Wang Y, Da Silva N, Kim B, Diao RY, Hill E, Brown D, Chan HC, Breton S (2014) CFTR interacts with ZO-1 to regulate tight junction assembly and epithelial differentiation through the ZONAB pathway. J Cell Sci 127(20):4396–4408

Ryeom SW, Paul D, Goodenough DA (2000) Truncation mutants of the tight junction protein ZO-1 disrupt corneal epithelial cell morphology. Mol Biol Cell 11(5):1687–1696

Saitoh Y, Suzuki H, Tani K, Nishikawa K, Irie K, Ogura Y, Tamura A, Tsukita S, Fujiyoshi Y, Tight junctions. (2015) Structural insight into tight junction disassembly by Clostridium perfringens enterotoxin. Science 347(6223):775–778

Saitou M, Fujimoto K, Doi Y, Itoh M, Fujimoto T, Furuse M, Takano H, Noda T, Tsukita S (1998) Occludin-deficient embryonic stem cells can differentiate into polarized epithelial cells bearing tight junctions. J Cell Biol 141(2):397–408

Sakurai A, Fukuhara S, Yamagishi A, Sako K, Kamioka Y, Masuda M, Nakaoka Y, Mochizuki N (2006) MAGI-1 is required for Rap1 activation upon cell-cell contact and for enhancement of vascular endothelial cadherin-mediated cell adhesion. Mol Biol Cell 17(2):966–976

Samanta P, Wang Y, Fuladi S, Zou J, Li Y, Shen L, Weber C, Khalili-Araghi F (2018) Molecular determination of claudin-15 organization and channel selectivity. J Gen Physiol 150(7): 949–968

Scott KG, Meddings JB, Kirk DR, Lees-Miller SP, Buret AG (2002) Intestinal infection with Giardia spp. reduces epithelial barrier function in a myosin light chain kinase-dependent fashion. Gastroenterology 123(4):1179–1190

Severson EA, Lee WY, Capaldo CT, Nusrat A, Parkos CA (2009) Junctional adhesion molecule A interacts with Afadin and PDZ-GEF2 to activate Rap1A, regulate beta1 integrin levels, and enhance cell migration. Mol Biol Cell 20(7):1916–1925

Shen L, Weber CR, Turner JR (2008) The tight junction protein complex undergoes rapid and continuous molecular remodeling at steady state. J Cell Biol 181(4):683–695

Shen L, Weber CR, Raleigh DR, Yu D, Turner JR (2011) Tight junction pore and leak pathways: a dynamic duo. Annu Rev Physiol 73:283–309

Sjö A, Magnusson KE, Peterson KH (2010) Protein kinase C activation has distinct effects on the localization, phosphorylation and detergent solubility of the claudin protein family in tight and leaky epithelial cells. J Membr Biol 236(2):181–189

Stevenson BR, Siliciano JD, Mooseker MS, Goodenough DA (1986) Identification of ZO-1: a high molecular weight polypeptide associated with the tight junction (zonula occludens) in a variety of epithelia. J Cell Biol 103(3):755–766

Takagaki Y, Manley JL (2000) Complex protein interactions within the human polyadenylation machinery identify a novel component. Mol Cell Biol 20(5):1515–1525

Takahisa M, Togashi S, Suzuki T, Kobayashi M, Murayama A, Kondo K, Miyake T, Ueda R (1996) The Drosophila tamou gene, a component of the activating pathway of extramacrochaetae expression, encodes a protein homologous to mammalian cell-cell junction-associated protein ZO-1. Genes Dev 10(14):1783–1795

Tsukita S, Furuse M (1999) Occludin and claudins in tight-junction strands: leading or supporting players? Trends Cell Biol 9(7):268–273

Tsukita S, Itoh M, Nagafuchi A, Yonemura S, Tsukita S (1993) Submembranous junctional plaque proteins include potential tumor suppressor molecules. J Cell Biol 123(5):1049–1053

Vianna JGP, Simor TG, Senna P, De Bortoli MR, Costalonga EF, Seguro AC, Luchi WM (2019) Atypical presentation of familial hypomagnesemia with hypercalciuria and nephrocalcinosis in a patient with a new claudin-16 gene mutation. Clin Nephrol Case Stud 7:27–34

Vietor I, Bader T, Paiha K, Huber LA (2001) Perturbation of the tight junction permeability barrier by occludin loop peptides activates beta-catenin/TCF/LEF-mediated transcription. EMBO Rep 2(4):306–312

Wang Z, Mandell KJ, Parkos CA, Mrsny RJ, Nusrat A (2005) The second loop of occludin is required for suppression of Raf1-induced tumor growth. Oncogene 24(27):4412–4420

Wang Z, Wade P, Mandell KJ, Akyildiz A, Parkos CA, Mrsny RJ, Nusrat A (2007) Raf 1 represses expression of the tight junction protein occludin via activation of the zinc-finger transcription factor slug. Oncogene 26(8):1222–1230

Wang SB, Xu T, Peng S, Singh D, Ghiassi-Nejad M, Adelman RA, Rizzolo LJ (2019) Disease-associated mutations of claudin-19 disrupt retinal neurogenesis and visual function. Commun Biol 2:113

Werth M, Walentin K, Aue A, Schönheit J, Wuebken A, Pode-Shakked N, Vilianovitch L, Erdmann B, Dekel B, Bader M et al (2010) The transcription factor grainyhead-like 2 regulates the molecular composition of the epithelial apical junctional complex. Development 137(22): 3835–3845

Willott E, Balda MS, Fanning AS, Jameson B, Van Itallie C, Anderson JM (1993) The tight junction protein ZO-1 is homologous to the Drosophila discs-large tumor suppressor protein of septate junctions. Proc Natl Acad Sci U S A 90(16):7834–7838

Wittchen ES, Haskins J, Stevenson BR (1999) Protein interactions at the tight junction. Actin has multiple binding partners, and ZO-1 forms independent complexes with ZO-2 and ZO-3. J Biol Chem 274(49):35179–35185

Wong V, Gumbiner BM (1997) A synthetic peptide corresponding to the extracellular domain of occludin perturbs the tight junction permeability barrier. J Cell Biol 136(2):399–409

Woo PL, Ching D, Guan Y, Firestone GL (1999) Requirement for Ras and phosphatidylinositol 3-kinase signaling uncouples the glucocorticoid-induced junctional organization and transepithelial electrical resistance in mammary tumor cells. J Biol Chem 274(46):32818–32828

Wu Y, Dowbenko D, Spencer S, Laura R, Lee J, Gu Q, Lasky LA (2000a) Interaction of the tumor suppressor PTEN/MMAC with a PDZ domain of MAGI3, a novel membrane-associated guanylate kinase. J Biol Chem 275(28):21477–21485

Wu X, Hepner K, Castelino-Prabhu S, Do D, Kaye MB, Yuan XJ, Wood J, Ross C, Sawyers CL, Whang YE (2000b) Evidence for regulation of the PTEN tumor suppressor by a membrane-localized multi-PDZ domain containing scaffold protein MAGI-2. Proc Natl Acad Sci U S A 97(8):4233–4238

Yamauchi K, Rai T, Kobayashi K, Sohara E, Suzuki T, Itoh T, Suda S, Hayama A, Sasaki S, Uchida S (2004) Disease-causing mutant WNK4 increases paracellular chloride permeability and phosphorylates claudins. Proc Natl Acad Sci U S A 101(13):4690–4694

Yao R, Natsume Y, Noda T (2004) MAGI-3 is involved in the regulation of the JNK signaling pathway as a scaffold protein for frizzled and Ltap. Oncogene 23(36):6023–6030

Yuhan R, Koutsouris A, Savkovic SD, Hecht G (1997) Enteropathogenic Escherichia coli-induced myosin light chain phosphorylation alters intestinal epithelial permeability. Gastroenterology 113(6):1873–1882

Zhu H, Liu Z, Huang Y, Zhang C, Li G, Liu W (2015) Biochemical and structural characterization of MUPP1-PDZ4 domain from Mus musculus. Acta Biochim Biophys Sin 47(3):199–206

Zihni C, Mills C, Matter K, Balda MS (2016) Tight junctions: from simple barriers to multifunctional molecular gates. Nat Rev Mol Cell Biol 17(9):564–580

The Role of Tight Junction Proteins in Cancer

8

Jayaprakash Narayana Kolla and Magesh Muthu

Abstract

Tight junctions (TJs) proteins are the network of proteins that play a major role in cell–cell communication and maintaining cell polarity. TJs primarily regulate the permeability of epithelial and endothelial cells, where it creates an intercellular barrier and paracellular diffusion fence. Apart from cell–cell communication functions, TJ proteins are involved in multiple intracellular signaling events in regulating cellular homeostasis. It is becoming evident that dysregulation of TJs results in various abnormalities including tumor progression and metastasis. In this chapter, we will discuss the basics and importance of TJs as well as their dysregulation in various cancer development. As it is apparent that TJs could be a potential therapeutic target in various tumor progression, we will also discuss various strategies for devising effective cancer-targeted therapy.

Keywords

Tight junction proteins · Claudins · Junctional complexes · Cancer · Epithelial–mesenchymal transition · Drug delivery · Signaling hubs

J. N. Kolla
Institute of Molecular Genetics of the Czech Academy of Sciences, Prague, Czech Republic

M. Muthu (✉)
Department of Oncology, John D. Dingell Veterans Affairs Medical Center, Barbara Ann Karmanos Cancer Institute, Wayne State University, Detroit, MI, USA

A. A. Bhat et al. (eds.), *Tight Junctions in Inflammation and Cancer*,
https://doi.org/10.1007/978-981-99-2415-8_8

8.1 Introduction

In multicellular organisms, both epithelial and endothelial cells are highly organized in a complex manner from lining the external surface to internal organs/glands to protect themselves and function physiologically to maintain homeostasis. For instance, the external epithelial cells lining the skin protect the organism against foreign substances like allergens and chemicals that are harmful to normal function, and the internal epithelial cells lining internal organs as well as glands in consensus with endothelial cells to protect against harmful pathogens from their external environment. Apart from protection, both epithelial and endothelial cells are in conjunction or communicated systematically in exchanging fluids of different molecular compositions including blood, urine, digestive juice, etc., for their physiological function in maintaining homeostasis.

Epithelial and endothelial cells are dynamically compartmentalized and facilitate interactions with the help of junctional complexes including tight junctions (TJs), adherens junctions (AJs), desmosomes, and gap junctions (Farquhar and Palade 1963) (Fig. 8.1). In brief, adherens junctions play a major role in the initiation and stabilization of cell–cell adhesion, intracellular signaling, and transcriptional regulation (Delva et al. 2009). Desmosome is also known as macula adherens, mainly involved in tethering arbitrate filaments to the plasma membrane. TJs require a high order of assembly of different channels of proteins that permit the exchange of ions, secondary messengers, and small molecular metabolites within adjacent cells

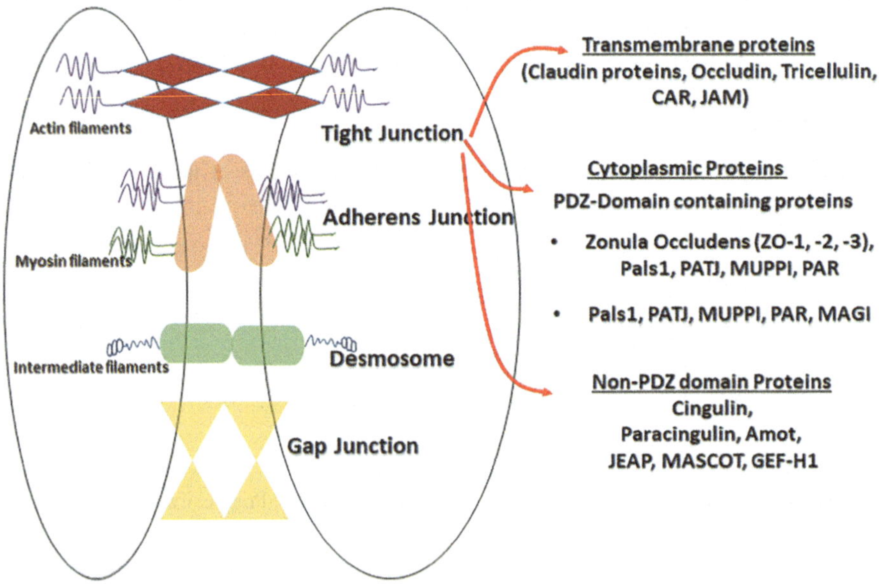

Fig. 8.1 Epithelial cell–cell junctional complexes. Schematic drawing of all the junctional components, where TJ proteins are located in the apical region of the cell

(Goodenough and Paul 2009). TJs, also known as zonula occludens (ZO), form protective barriers that are mainly involved in a selective exchange of ions, water, and other molecules via a paracellular pathway to maintain cell polarity. TJs dysregulation leads to the alteration of permeability barrier function leading to the cause of various diseases including cancer. For instance, cancer cell requires invasion into neighboring cells by breaching protective barriers such as cell junctions which eventually leads to metastasis. In this section, we will focus mainly on the basics of tight junctions, and how their dysregulation contributes to various types of cancer development as well as TJs as a potential therapeutic target in cancer therapy.

8.2 Tight Junction Proteins and Function

Tight junctions composed of multilayer components: (1) transmembrane proteins including occludin, claudins, tricellulin, junctional adhesion molecules (JAM), and coxsackievirus and adenovirus receptor (CAR) which mainly constitutes TJs strands; (2) cytoplasmic proteins broadly classified as zona occludens (ZO-1, 2, 3), proteins connected with Seven Lin 1 (Pals1), Pals1 associated TJ protein (PATJ), multi-PDZ domain protein 1 (MUPP1), partitioning defective (PAR), MAGI (MAGUK with inverted subdomain structure)—sub-family of MAGUK (membrane-associated guanylate kinases) and non-PDZ proteins that collectively facilitate the interaction of integral transmembrane proteins with actin filaments and other cytoplasmic proteins that result in a cascade of events including tight junction transcription and regulation (Itoh and Bissell 2003; Salvador et al. 2016; Guillemot et al. 2008).

8.2.1 Transmembrane Proteins

8.2.1.1 Occludin

It is the 1st TJ protein to be discovered at the molecular level and identified as ~65 kDa protein which is located at TJ both in endothelial and epithelial cells (Furuse et al. 1993). The initial discovery of the occludin suggested that it is an essential component to mediate the protective barrier function of TJ, subsequently, Saitou et al. (2000) demonstrated that occludin knockout mice do not show any structural alteration in TJ development. However, postnatal growth retardation was observed in occludin knockout mice in comparison to wild-type mice, i.e. histological staining of various tissues revealed significant abnormalities at 3–6 weeks old knockout mice displayed loss of differentiation of epithelial cells in the gastric gland and 28–60 weeks old mice displayed chronic gastritis, brain calcification, and further irregularities in testis, salivary gland as well as bone (Saitou et al. 2000). Overall, occludin is indispensable at the cellular level, though more detailed studies are needed to see the complexity of its regulation and function.

8.2.1.2 Claudins

Claudins are a large group of transmembrane proteins that are considered the molecular backbone of TJ. Similar to occludin, claudins are also transmembrane domain proteins of ~23 kDa and at least there are 27 claudins reported to date. Further, it is well documented that claudins make the core of TJ complex (Cording et al. 2013), and several other studies also revealed that the family of claudin proteins critically regulates physiological conditions in vivo in a tissue-specific manner. For instance, claudin-5 is majorly expressed in brain endothelial cells to form the blood–brain barrier (Morita et al. 1999), which controls blood circulation in the brain and prevents the entry of toxic substances into the brain. Similarly, another member of the claudin family, claudin-11 is highly expressed in Sertoli cells, and loss of its function leads to alteration in the blood–testis barrier eventually resulting in infertility (Gow et al. 1999), which suggests that the claudin family maintains the integrity of TJ in controlling the physiological plasticity to maintain cellular homeostasis. Additionally, the family of claudin proteins is dysregulated at transcriptional levels resulting in various diseases like cancer (Morin 2005), for example, downregulation of claudin-1, -7 and upregulation of claudin-3, 4 contributes to diverse stages of breast cancer (Kramer et al. 2000; Kominsky et al. 2003, 2004). To better understand how different members of the claudin family contribute to various cancer development in detail will be focused on in the latter part of this chapter.

8.2.1.3 Tricellulin

It is also known as MARVELD2 (MARVEL Domain Containing 2), which regulates the configuration of cell–cell junctions in epithelial cells (Oda et al. 2014). Tricellulin is highly localized at tricellular contacts, which increases the affinity of adhesion between epithelial cells (Morampudi et al. 2016). Recent studies reported that homozygous mutation in tricellulin coding gene MARVELD2 (DFNB49) leads to deafness (Mašindová et al. 2015), while loss of function of tricellulin results in structural aberration of epithelial cells.

8.2.1.4 Coxsackievirus and Adenovirus Receptor (CAR)

CAR is a ~40 kDa integral membrane protein member of TJ, which is highly localized in extracellular space and serves as a receptor for coxsackie and adenoviruses (Reeh et al. 2013). Unlike claudin and occludin, it is a single transmembrane protein and several studies revealed that the deregulation of CAR contributes to inflammation and various types of cancer development.

8.2.1.5 Junctional Adhesion Molecule (JAM)

JAM is a ~43 kDa protein, which belongs to the immunoglobulin superfamily of cell adhesion receptors, JAM-A/1 was the first protein to be discovered in this family (Kornecki et al. 1990; Martìn-Padura et al. 1998) and eventually, other members were identified, namely JAM-B/2, JAM-C/3 and distantly related JAM-L (Aurrand-Lions et al. 2001; Arrate et al. 2001; Luissint et al. 2008; Hirabayashi et al. 2003; Mori et al. 2004). It is localized at TJs of many cells including epithelial and endothelial cells as well as in leukocytes and platelets. Similar to CAR, JAM is also a single transmembrane protein, which interacts with several scaffolding proteins to regulate cell polarity, angiogenesis, and cellular morphology. For example, JAM-2 and -3 are highly colocalized with PAR-3 to maintain cell polarity (Brennan et al. 2010; Ebnet et al. 2003). Similarly, the interaction of JAM-2 and -3 regulates the polarity of the blood–testis barrier (Ebnet et al. 2004) and loss of JAM-3 activations has been shown to affect fertility by blocking the germ cell's growth and development (Ebnet et al. 2017).

8.2.2 Cytoplasmic Proteins

8.2.2.1 PDZ-Domain Containing Proteins

In TJ, PDZ-domain-containing proteins are mainly involved in the gathering of TJ strands and facilitate the anchoring of membrane and other signaling proteins that are postulated in the structural organization and expression of genes.

8.2.2.2 Zonula Occludens (ZO-1, ZO-2, and ZO-3)

ZO-1 (~220 kDa) is a cytoplasmic scaffolding protein colocalized with TJ in various epithelial cells (Gumbiner et al. 1991). ZO-2 (~160 kDa) is identified as a binding partner of ZO-1 in maintaining cell polarity and stability. ZO-3 is also known as tight junction protein 3, unlike ZO-1 and ZO-2, it is present in site-specific epithelial TJ, and not in endothelial or cadherin-based cell–cell adhesion sites (for example, adherens junctions of fibroblasts) (Inoko et al. 2003). Apart from scaffolding functions, ZO proteins can shuttle within the nucleus and TJ which explains the dynamics of TJ in regulating basic processes like cell proliferation and differentiation. To further understand the significance of ZO proteins, recent findings revealed that the knockout of ZO-1 and -2 resulted in the lethality of embryos by extreme abnormal apoptosis. This study further suggests that embryonic lethality is highly possible due to an alteration in TJ regulation and its protective barrier function (Katsuno et al. 2008), while deregulation of ZO-3 results in impaired TJ assembly and its function.

8.2.2.3 Pals1, PATJ, MUPP1, PAR

Pals1 is a subset of MAGUK (membrane-associated guanylate kinase) proteins and comprises a PDZ domain that attaches to the carboxy-end of another transmembrane protein CRB3 (human orthologue of Drosophila crumbs protein). Similarly, Pals1 also binds to PATJ, where it acts as an adapter protein between CRB3 and PATJ (Makarova et al. 2003; Lemmers et al. 2002). Further analysis revealed that PATJ binds to PDZ domains of other TJs such as claudin-1 and ZO-3 (Roh et al. 2002). MUPP1 is a multi-PDZ domain protein known to associate with claudin-4 and -8 (Jeansonne et al. 2003).

PAR is a conserved protein involved in polarization and asymmetric cell division. PAR-3 is known to localize with aPKC and other proteins at the tight junction in determining cell function by regulating mitotic spindle coordination in the cell cycle (Kemphues 2000). In addition to aPKC, PAR-3 colocalizes with PAR-6 (another member of the PAR family), and the whole complex, in turn, binds directly to JAM-1 (Qiu et al. 2000; Joberty et al. 2000). Overexpression of Par-3 has been shown to affect the formation of TJ in the initial phase of its development, which clearly emphasizes the importance of the PAR protein complex in the polarization epithelial cell.

8.2.2.4 MAGI

MAGI is a multi-domain protein that contains PDZ, the Src homology 3 (SH3), and the guanylate kinase (GUK) domains. It has been observed that MAGI-1 may have similar characteristics to ZO-1 as it can shuttle between the nucleus and TJ, though is yet to understand its definitive function in TJs (Dobrosotskaya et al. 1997).

8.2.2.5 Non-PDZ Proteins

Non-PDZ proteins are a family of cytoplasmic proteins involved in scaffolding TJs such as cingulin, paracingulin, Amot, JEAP, MASCOT, GEF-H1, and symplekin. Most of the members of this family have been shown to regulate TJs indirectly, yet more investigation is needed to understand their interaction with TJ proteins and functions (Guillemot et al. 2008).

8.3 TJ Proteins in Cancer

In humans, every cell has to undergo a life-and-death process to rejuvenate itself, and maintain cell plasticity, and overall cellular homeostasis. Unfortunately, some cells fail to follow this process, i.e. these cells compromise the process called apoptosis (programmed cell death), which results in the shift of normal cells to cancer cells. These cells grow abnormally and have the potential to invade other parts of the body. Over the years, several studies have reported that various factors impart to the cancer cells invading distant sites, one such key factor is the dysregulation of cellular junctions, specifically members of TJ proteins that play a decisive role in various metastatic cancer progression.

8.3.1 The Role of TJs in Various Cancers

As we discussed above TJs are important in cell–cell communication to regulate the selective exchange of information across epithelium (Brucher and Jamall 2014). Given the fact that TJs play an important role in the proliferation of cells and also in cell differentiation, it is obvious that derailed TJ regulation lends to a favorable environment for tumor progression. Here we will focus on how TJs and their associated proteins contribute to various cancers and their microenvironment.

8.3.1.1 Breast Cancer

Several studies revealed TJs alteration mediates abnormal cellular polarity, for instance, claudin-1 is present in normal breast epithelial cells and it is either

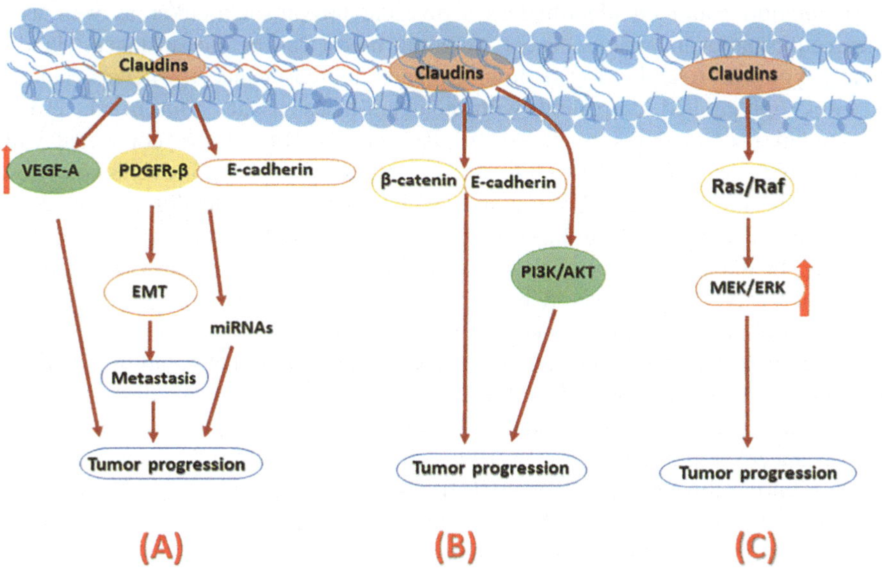

Fig. 8.2 Mechanisms of claudins dysregulation in tumorigenesis. (**a**) Downregulation of claudin-5 results in the upregulation of VEGF-A in GBM tumor progression; overexpression of claudin-1 deregulates EMT-related genes including PDGFRβ, cadherin as well ascertains miRNAs in breast cancer progression. (**b**) Dysregulation of claudin-1 induces β-catenin as well as cadherin expression in colon cancer as well as modulating PI3K/AKT pathway in prostate cancer. (**c**) Overexpression of claudin-2 activates the Ras/Raf/MEK/ERK pathway in osteosarcoma cells

downregulated or completely absent in human breast cancer cell lines including MDA-MB-361 and MDA-MB-435 (Hoevel et al. 2002). Further, re-introducing claudin-1 in these cells reverses the permeability of the cell membrane and its function of paracellular diffusion, suggesting that it could play a potential tumor suppressor role, and loss of its function leads to breast cancer development. In another study, overexpression of claudin-1 results in the deregulation of a group of miRNAs (miR-9-5p, miR-9-3p, etc.) in the MDA-MB231. Interestingly, these miRNAs have been highly linked with tumor suppression in breast cancer. Moreover, overexpression of claudin is correlated with the deregulation of epithelial–mesenchymal transition (EMT) connected genes PDGFRβ (platelet-derived growth factor receptor β) and E-cadherin. Collectively it is clear that claudin-1 dysregulation influences the functions of miRNAs that are involved in breast cancer progression (Fig. 8.2). Similarly, overexpression of claudins -2, -3, -4, and -20 in breast cancer has been shown to promote liver metastasis by forming new vascular channels, reducing apoptosis, and displaying aggressive phenotype (Majer et al. 2016; Tabariès et al. 2012; Martin et al. 2013). Meanwhile, a recent report revealed the relationship between claudin and interleukin (IL)-18 by enhancing the migratory ability of the MCF-7. In this study, it has been shown that IL-18 enhances cell migration by downregulating claudin-12 and triggering p38 MAPK (mitogen-activated protein kinase) pathway. Additionally, several studies indicate that claudin dysregulation could be identified as a possible prognostic marker as well as a therapeutic target in various cancer including breast cancer. For instance, claudin family types -2, -6, and -14 were downregulated in breast cancers, whereas claudin-11 was upregulated compared with non-cancerous tissues (Jia et al. 2019). Upregulation of claudins-2, -5, -6, -9, -10, -11, and 14–20 was connected with relapse-free survival (RFS), while higher levels of claudin-3 expression were associated with poor RFS. Yang et al. (2021) reported that claudin-3, -5, and -11 were not only prognostic indicators but also therapeutic targets in breast cancer patients. In the majority of triple-negative breast cancers, claudin-3 and -7 were aberrantly expressed and could be the prominent prognostic markers as well as indicators of poor survival (Jääskeläinen et al. 2018).

In addition to claudins, ZO-1/ZO-2 is another TJ protein that could be a possible tumor suppressor in regulating breast epithelial cells. Recent findings noted decreased expression of ZO-1 and ZO-2 in several cancer cell lines and human breast cancer samples (Hoover et al. 1998; Chlenski et al. 2000). In contrast, another study reported that upregulation of ZO-1 resulted in the activation of insulin-like growth factor I receptor (IGF-IR), which in turn enhanced E-cadherin-induced cell–cell adhesion to blunt the cancer cell invasive ability (Mauro et al. 2001). Similarly, ZO-2 overexpression leads to downregulation of the adenovirus type 9 oncogenic determinant E4 protein, which is a unique promoter of breast cancer development (Glaunsinger et al. 2001).

JAM proteins are depicted to control epithelial cell adhesion, and phenotype via enhancing β1-integrin expression by Rap1 GTPase and it is obvious altering JAM function could lead to distribution in cell membrane integrity and polarity to promote cancer progression. Indeed, it has been found that JAM-A protein expression is lower in highly migratory breast cancer cells including MDA-MB-468 and MDA-MB-231. Furthermore, overexpression of JAM-A protein expression in MDAMB-231 inhibited migration and invasion in collagen culture (Naik et al. 2008).

Occludin is another tight junction protein indicated to be reduced in breast cancer and its re-expression resulted in the deregulation of anti-apoptosis-associated genes such as bcl-2 and survivin (Osanai et al. 2006, 2007) (Fig. 8.3).

8.3.1.2 Gliomas

Gliomas are tumors from brain cells. In brief, astrocytes are members of a glial family which plays a vital role in brain development via connecting neurons and other brain cells including endothelial and microglia cells. Although the mechanism is not yet clear, astrocytes undergo oncogenic transformation resulting in gliomas. Glioblastoma multiforme (GBM) is an aggressive form of a tumor with excessive vascularization and it originates from well-developed gliomas as a precursor. At present, it is very hard to treat glioblastoma due to the presence of the blood–brain barrier (BBB), where tumor cells are protected by the stable environment by the extended blood vessels, thereby they can colonize the adjacent healthier cells and compromise the drug delivery across BBB (Dubois et al. 2014).

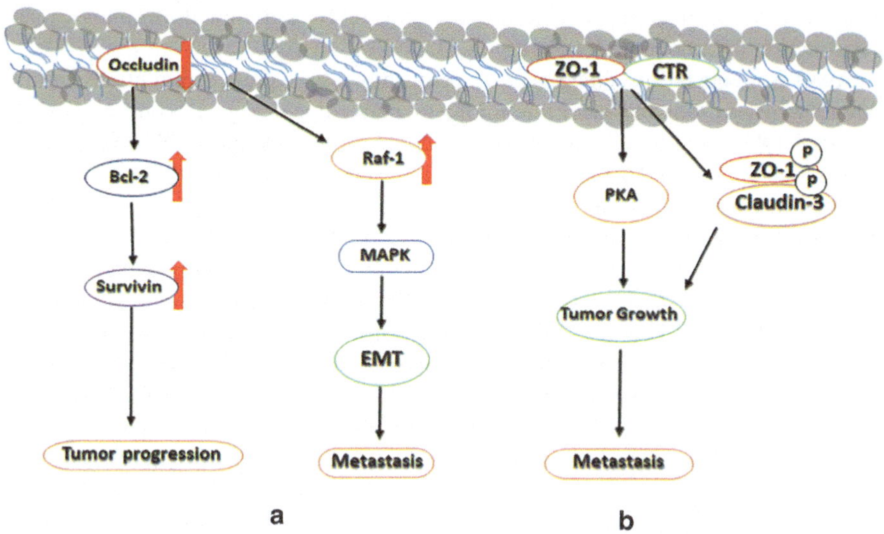

Fig. 8.3 Role of occludins and ZO-1 in tumorigenesis. (**a**) Downregulation of occludin results in upregulation of anti-apoptotic genes, Bcl-2, and survivin in promoting breast cancer; similarly, downregulation of occludin results in upregulation of oncogene Raf-1, an activator of MAPK pathway, in turn, mediates lung cancer metastasis (**b**) Activated calcitonin receptor (CTR) interacts with ZO-1, results in activation of cAMP protein kinase (PKA) as well as phosphorylation of claudin-1 and ZO-1 leads to prostate cancer and metastasis

An attempt to screen TJs protein in tumor tissues from GBM's patients revealed that compared to microvessels, hyperplastic vessels showed downregulation of occludin and complete absence of claudin-1 in almost all tumor microvessels. It is also reported that downregulation of occludin and claudin-5 mediates upregulation of vascular endothelial growth factor-A (VEGF-A) to promote angiogenesis and cell proliferation (Fig. 8.2). Thus these studies suggest that dysregulation of TJs proteins alters cell morphology and permeability to support GBM tumor progression (Argaw et al. 2009; Salvador et al. 2015).

8.3.1.3 Lung Cancer

TJs of the lung or pulmonary epithelial cells provide structural stability to regulate the air–blood barrier to allow selective solutes or ions into the lungs and also prevent the ingestion of pathogenic microorganisms or substances. Immunohistochemical staining of patient tissue samples of various lung cancer subtypes revealed that occludin expression is completely absent in squamous cell carcinomas (SQCC), small cell carcinomas (SCC), large cell carcinomas (LCC), and adenocarcinomas (AC) (Tobioka et al. 2004). Similarly, another report exposed that the downregulation of occludin increases the expression of oncogene Raf-1, which is a key regulator of the MAPK pathway to promote cell proliferation and differentiation. Further Raf-1 induces an epithelial–mesenchymal transition to promote metastasis, while restoring occludin expression rescues epithelial morphology and TJs function (Wang et al. 2005) (Fig. 8.3).

Apart from occludins, claudins have been shown to express differentially in lung cancer subtypes. For instance, immunohistochemical staining of primary epithelial tumors (SQCC, SCC, AC) showed higher expression of claudin-1. Similarly higher expression of claudin-2, -3 and -4 in SQCC as well as in SCC were observed, where as clauidn-5 was weakly expressed in AC (Soini 2012). Interestingly, expression levels of the claudin family such as claudin -1, -3, -4, and -7, and ZO-2 and -3 were downregulated in lung adenocarcinoma compared with normal bronchial cells (Soini 2012). It indicates that the depiction of TJs in lung cancers can be valuable prognostic biomarkers and offer new perceptions of their histogenesis. Differential expression of claudin-1 for small cell lung cancer (SCLC) and claudin-6 for non-small cell lung cancer (NSCLC) was reported as significant prognostic biomarkers (Soini 2011). Additionally, claudin-6 was an important diagnostic marker due to its lower expression levels which are highly correlated with poor survival rate in NSCLC (Wang et al. 2015). Wada et al. (2013) described that CLDN15 can be used in the clinical setting as a positive marker for malignant pleural mesothelioma diagnosis. Additionally, Eguchi et al. (2021) reported that the abnormality of claudin-2 expression in lung adenocarcinoma tissues could lead to drug resistance.

8.3.1.4 Colorectal Cancer

In the intestine, claudins play a key role in the selective transport of Na^+, which is critical in absorbing essential nutrients to support cell growth (Wada et al. 2013). Claudin-1 induces cellular transformation and metastasis by mediating β-catenin signaling and E-cadherin expression in colon cancer (Dhawan et al. 2005). Meanwhile, JAM-A expression is higher in the human gastric cancer cell line (NCI-N87), where knockdown of JAM-A significantly inhibits cell proliferation by modulating anti-apoptotic protein Bcl-xL, suggesting a vital role of JAM-A in gastric cancer progression (Ikeo et al. 2015). Interestingly, the expression pattern of TJ proteins occludin, claudin-1, -4, and ZO-1 in advanced colon cancer revealed their prognostic worth. High-grade colonic tumors were diagnosed with low expression levels of claudin-1 and ZO-1 and further loss of claudin-1 was the indicator for recurrence of the disease and poor survival rate (Resnick et al. 2005). The differential expression pattern of claudin-4 in metastatic lesions of colorectal cancers implicated its role as an important biomarker for detecting the danger of upstage metastasis (Shang et al. 2012). Colorectal cancers are frequently encountered with downregulation of claudin-8 and contrary upregulation of claudin-1 and -12 (Zhu et al. 2019). The majority of colorectal cancers overexpress claudin-1 which may signify a target molecule for cancer treatment apart from prognosis. Kaihara et al. (2003) described that ZO-1 was overexpressed in normal epithelial cells of the colorectum and reduced in primary colorectal cancer, further upregulated in metastatic cancers. Tabariès et al. (2021) indicated that claudin-2 levels in extracellular vesicles from patients may help as an appropriate prognostic marker to envisage the metastatic development from colorectal cancer to the liver.

8.3.1.5 Prostate Cancer

A plethora of studies have shown that alteration in TJs leads to prostate cancer. While flavonoids extracted from *Orostachys japonicus* A. Berger (FEOJ) mediate suppression of claudin-1, -3 through downregulation of PI3K/AKT pathway, thereby inhibiting the invasive and migratory ability of prostate cancer cells (Shin et al. 2013). Furthermore, it has been shown that hepatocyte growth factor (HGF) regulates TJs in prostate epithelial cells and it is also observed that HGF could influence metastasis of prostate cancer. Meanwhile, calcitonin receptor (CTR) is a member of G-protein coupled receptors (GPCRs) and plays a vital role in regulating calcium homeostasis and it has been shown that expression of calcitonin and its receptor (CTR) is upregulated in metastatic prostate cancer (Chien et al. 2001). Additionally, it is revealed that activated CTR interacts with ZO-1 and this interaction, in turn, mediates activation of cAMP protein kinase (PKA) as well as phosphorylation of claudin-3 and ZO-1, collectively these cases lead to CTR accelerated tumor growth and metastasis (Aljameeli et al. 2017).

In a study to screen healthy samples of prostate was shown that the predominant expression of a shortened form of claudin-7 and downregulation of claudin-1, -3, and -7 was associated with advanced cancer stages and relapses of the disease, whereas claudin-4 was confined only to prominent tumor grade (Seo et al. 2010). Landers et al. (2008) studied the levels of claudin-3 and -4 in normal human prostate samples and described them as possible targets for Clostridium perfringens enterotoxin (CPE) interceded treatment for prostate cancer. The tenacious increased level of claudin-3 in prostate cancer and efficient cytotoxicity of CPE in metastatic androgen-independent prostate cancer proposes a novel possible therapeutic approach for prostate cancer.

8.3.1.6 Bladder Cancer

Boireau et al. (2007) and Törzsök et al. (2011) investigated the expression pattern of claudins-1, -4, and -7 in bladder carcinoma and found that claudin-4 was notably altered in many tumors contrary to rare changes noticed in the expression of claudins-1 and 7. Overexpressed claudin-4 was tend to be downregulated in highly invasive tumors, and this form was linked with limited period survival of bladder tumor patients. Xu et al. (2019) described that ZO-1 allied nucleic acid adhering protein is overexpressed in bladder cancer cell lines, which enhances invasion, representing the pathological significance in the early diagnosis of gall bladder tumors.

8.3.1.7 Skin Cancer

Heterogeneous expression of claudins and other TJs played a significant role in skin cancer progression. By using tissue microarray technology, Leotlela et al. (2007) exposed that claudin-1 was upregulated in melanoma and irrationally uttered in the cytoplasm of malignant cells. The expression of claudin-11 has been proposed as a prognostic marker for later-stage cutaneous squamous carcinoma and meditates the distinct stages of tumor progression and differentiation (Shen et al. 2017).

8.3.1.8 Esophageal Cancer

In the majority of esophageal cancers, the claudin family significantly provided prognostic indicators in the early diagnosis of tumor development. Miyamoto et al. (2008) suggested that the expression of claudin-1 is connected with the relapse grade and miserable prognosis in esophageal cancer and its appearance may be a valuable tool in the prediction of recurrence in esophageal cancer. Highly invasive and metastatic esophageal cancers were significantly connected with the downregulation of claudin-7 and associated with tumor progression (Usami et al. 2006; Lioni et al. 2007). The loss of occludin in esophageal squamous cell carcinoma was also a significant prognostic marker (Qin et al. 2017). An immunohistochemical microarray study in esophageal squamous cell carcinoma patients revealed that claudin-1 and -4 were significantly induced in cancer tissues than in normal tissue. Additionally, claudin-4 was an important biomarker for the indication of poor survival in esophageal squamous cell carcinoma (Liu et al. 2021).

8.3.1.9 Gastric Cancer

Heterogeneous expression of claudins-1, 3, and 4 besides ZO-1 highly correlated in a majority of patients with gastric cancer, not only a useful prognostic tool in early diagnosis but also an indicator of poor survival condition (Resnick et al. 2005). TJs were sturdily articulated in the majority of gastric adenocarcinomas nonetheless less often in circulated gastric cancers. The overexpression of claudin during gastric cancer progression suggests their likely utility as prognostic markers and potential targets for cancer treatment. Jun et al. (2014) suggested that the overexpression of claudin-3 and 7 and the loss of claudin-18 controlled the carcinogenesis of gastric cancer. Immunohistochemically, diverse expressions of claudin-3, -4, and -18 were noticed in very advanced stages of gastric cancer. Moreover, claudin-4 was an important biomarker for gastric cancer patient prognosis and is valuable in the organization of gastric cancer (Liu and Li 2020; Li et al. 2020). The overexpression of claudin-7 was more critical in gastric adenocarcinoma than in normal gastric epithelium, dramatically losing in diffuse-type gastric adenocarcinoma (Park et al. 2007).

8.3.1.10 Gynecological Cancers

Claudins-1, -2, -4, and -7 were found in the intact epithelium in cervical intraepithelial neoplasia (CIN), displaying a reduction in carcinoma in situ (CIS) (Sobel et al. 2005). In the early stages of cervical carcinogenesis, occludin and claudin-2 were overexpressed, whereas claudin-1, -4, and -7 were restricted to only normal epithelia (Zinner et al. 2013). In ovarian carcinoma, claudin-7 was highly differentiated and upregulated in early and metastatic tumors (Dahiya et al. 2011). From the previous reports, it is evident that both claudins-3 and -4 vary in expression during ovarian cancer progression and are useful markers for ovarian tumors (Litkouhi et al. 2007). Claudin-4 is upregulated in the majority of ovarian cancer, which may lead to novel approaches for therapy and early diagnosis of disease (Hicks et al. 2016). Nissi et al. (2015) described that overexpression of claudin-5 was associated with advanced ovarian adenocarcinoma. Additionally, occludin correlated with lower expression in metastatic stages.

8.3.1.11 Bone Cancer

Martin et al. (2016) demonstrated that the expression of occludin has a strong association with bone metastasis in human cancer. Claudin-2 exhibited tumor suppressor function which was expressed typically in normal osteoblast cells but downregulated in osteosarcoma cells; these expression levels have the potential to be useful as molecular tools for the diagnosis of osteosarcoma (Zhang et al. 2018).

8.4 Role of TJs in Drug Discovery

It is conceivable that dysregulated TJs are well studied and could be useful to identify differentiated adenocarcinomas, possible prognostic markers, and therapeutic targets. Variable expression of TJs may redirect the different stages of tumor progression and implicate them as prognostic tools for early diagnosis of tumor stages. Furthermore targeting dysregulated TJs would be an effective strategy to develop novel therapies to encounter cancer development.

8.4.1 Drug Availability and Permeability

Drug availability and permeability were a big challenge in drug development involving several formulations, and modifications were major interests in the optimization of drug availability. However, the inflection of TJs is an efficient approach for the aggregate absorption of drugs, these TJ modulators could be dynamic excipients in drug formulations. Several studies revealed various methods of drug delivery mode, to devise an effective strategy to target TJs in various cancers.

8.4.1.1 Ocular Drug Delivery

TJs among corneal epithelial cells compromise resistance to hydrophilic molecules. The absorption of Ca^{2+} ions is vital for the modulation of these TJs (Morrison and Khutoryanskiy 2014). Calcium chelators are substances that binds and sequester calcium ions including polyaminocarboxylic acids such as EGTA, ethylenediamine-N, N'-disuccinic acid, EDTA, and are used in topical ocular formulations as stabilizing agents and infusion improvement via perturbation of TJs (Moiseev and Khutoryanskiy 2019)

8.4.1.2 Nasal Drug Delivery

A major hurdle in nasal drug delivery was the low absorptivity of high molecular weight drugs (Ozsoy et al. 2009). The C-terminal fragment of CPE (C-CPE) regulates the barrier function through claudins. CPE and Zonula occludens toxin (ZoT) from Vibrio cholera were reported as modulators of tight junctions such as claudin and ZO, employed to enhance the nasal absorption of large molecule drugs (Uchida et al. 2010; Gopalakrishnan et al. 2009).

8.4.1.3 Intestinal Drug Delivery

TJs such as claudins, occludin, and ZO are essential for the integrity of the epithelial barrier, modulation of these proteins may be helpful to enhance oral drug delivery (Maher et al. 2009). Optimization and formulation are inevitable for a variety of complex drugs, including proteins, peptides, viral vectors, and nucleic acids (Deli 2009). TJs modulator methotrexate was well studied for regulation of TJs and improved intestinal absorption of itself and other immunosuppressant or therapeutic drugs (Hamada et al. 2013).

8.4.1.4 Delivery Across the Blood–Brain Barrier

Based on their binding properties, TJ regulators for drug delivery are categorized into 1st generation, which contains toxins and their fragments, and 2nd generation binders included antibodies (Tachibana et al. 2020). Among the claudin family, claudin-5 was important for the integrity barrier at the blood–brain barrier (Hashimoto et al. 2019) and henceforth several models were established around the claudin-5 target to enhance drug absorptivity (Tachibana and Kondoh 2021).

8.4.2 TJs as Therapeutic Targets

TJ proteins emerged as not only important prognostic factors in cancer metastasis but also therapeutic targets in cancer. Zeisel et al. (2019) reviewed about TJ proteins are emerging as targets for novel therapeutic approaches for gastric and hepatic cancers. Du et al. (2021) reported that in various tumors such as esophageal, liver, endometrial, and ovarian cancers, research systematically shows that claudin-6 was expressed in tumor tissues but is not expressed or is expressed at low levels in encompassing normal tissues. In such tumors, claudin-6 could be a carcinoembryonic prognostic marker and also a drug target. In the antibody-dependent therapies of cancer, claudins were the significant target, the antibodies which developed against claudin-1, -3, and -4 were well studied (Offner et al. 2005). In colorectal cancers, the antibody developed against claudin-1 exhibited anti-metastatic potential and proved the reduction of cancer progression, the anti-claudin-4 antibody also observed as synergistic when combined with other therapeutic agents like 5-fluorouracil and anti-EGFR antibodies (Fujiwara-Tani et al. 2018). Interestingly, the monoclonal antibodies developed against claudin-6 and claudin-18.2 were advanced to clinical trials in the treatment of gastric cancer and ovarian cancer, respectively (Stadler et al. 2015; Liang et al. 2021), these antibodies were also proved to be clinically safe. A first-generation chimeric monoclonal antibody claudiximab (zolbetuximab or IMB362), which targets claudin-18.2 was well investigated in metastatic gastric cancers. It revolutionizes claudin-based monoclonal antibody therapy in cancers (Ungureanu et al. 2021). IMB362 activates the antibody-dependent cellular cytotoxicity and complement-dependent cytotoxicity

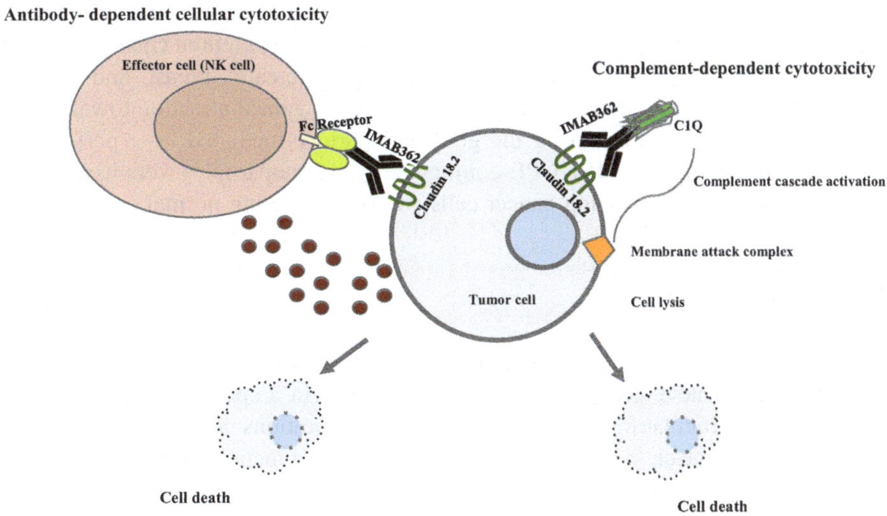

Fig. 8.4 Anticancer mechanism of IMB362 (claudiximab or zolbetuximab)

Table 8.1 Active clinical candidates targeting claudin-18.2

Cancer type	Drug (pipeline)	clinicaltrials.gov identifier
Advanced solid tumors	CAR-T cell therapy, CT041, LCAR-C18S, and TST001 (phase 1)	NCT03874897 NCT04404595 NCT04495296 NCT04467853
Pancreatic cancer	Zolbetuximab—IMAB362 (phase 2)	NCT03816163
Locally advanced unresectable or metastatic gastric or gastroesophageal junction (GEJ) adenocarcinoma	Zolbetuximab—IMAB362 (phase 3)	NCT03653507
Advanced solid tumors	SYSA1801, human anti-claudin-18.2 monoclonal antibody-MMAE drug conjugate (phase 1)	NCT05009966
Gastric and gastroesophageal junction adenocarcinoma	AMG 910, bispecific T-cell engager antibody (phase 1)	NCT04260191
Advanced solid tumors	Anti-CLDN18.2 ADC CMG901 (phase 1)	NCT04805307
Advanced solid tumors	AB011, human anti-claudin-18.2 monoclonal antibody	NCT04400383

through binding claudin 18.2, further inducing cell death of tumor cells (Fig. 8.4). The majority of actively recruiting clinical candidates are targeted claudin-18.2, including therapeutics such as chimeric antigen receptor (CAR) T-cell therapy and monoclonal antibodies (Table 8.1) (https://clinicaltrials.gov/). CPE binding to claudin attained importance in receptor distraction-based therapeutics in gastrointestinal cancers, especially claudin-3 and -4 played a role as CPE receptors (Li 2021). The binding capability of CPE to claudins has a potential chance to target claudin-3 and -4 dysregulated cancers, especially pancreatic, breast, and ovarian cancers. Recently, nanoparticle-based targeting cancer gained attention toward tight junctions like claudins. By using the gold nanoparticle-mediated laser perforation technique (Becker et al. 2019), CPE-conjugated gold nanoparticles were used to kill specifically claudin-expressing cancer cells without impacting normal cells.

8.5 Conclusion

TJs protein plays a pivotal role in paracellular transport and cell barrier function. Moreover, TJ protein expression in different tissues in a specific manner plays a unique role in regulating normal physiological conditions as well as abnormal conditions like cancer. Alteration of TJs structure and function eventually leads to abnormal cell differentiation and proliferation, ultimately resulting in various cancer development. From our review, it is understandable that TJ proteins are dynamic as they can be a potential tumor suppressor in normal cell growth as well as a prognostic marker in cancer cells which leaves a footprint on various stages of cancer development. The never-ending fight against cancer across the world is evident and we need to devise various effective strategies to inhibit tumor progression. It is convincing that various TJ proteins including claudins could play a significant role as a therapeutic target to develop novel drugs and vaccines to inhibit tumor progression in nearby future.

References

Aljameeli A, Thakkar A, Shah G (2017) Calcitonin receptor increases invasion of prostate cancer cells by recruiting zonula occludens-1 and promoting PKA-mediated TJ disassembly. Cell Signal 36:1–13. https://doi.org/10.1016/j.cellsig.2017.04.008

Argaw ATGB, Zhang Y, Zameer A, John GR (2009) VEGF-mediated disruption of endothelial CLN-5 promotes blood-brain barrier breakdown. Proc Natl Acad Sci U S A 106(6):1977–1982. https://doi.org/10.1073/pnas.0808698106

Arrate MP, Rodriguez JM, Tran TM, Brock TA, Cunningham SA (2001) Cloning of human junctional adhesion molecule 3 (JAM3) and its identification as the JAM2 counter-receptor. J Biol Chem 276:45826–45832. https://doi.org/10.1074/jbc.M105972200

Aurrand-Lions M, Duncan L, Ballestrem C, Imhof BA (2001) JAM-2, a novel immunoglobulin superfamily molecule, expressed by endothelial and lymphatic cells. J Biol Chem 276:2733–2741. https://doi.org/10.1074/jbc.M005458200

Becker A, Lehrich T, Kalies S, Heisterkamp A, Ngezahayo A (2019) Parameters for optoperforation-induced killing of cancer cells using gold nanoparticles functionalized with the C-terminal fragment of clostridium perfringens enterotoxin. Int J Mol Sci 20(17):4248. https://doi.org/10.3390/ijms20174248

Boireau S, Buchert M, Samuel MS, Pannequin J, Ryan JL, Choquet A, Chapuis H, Rebillard X, Avances C, Ernst M, Joubert D, Mottet N, Hollande F (2007) DNA-methylation-dependent alterations of claudin-4 expression in human bladder carcinoma. Carcinogenesis 28:246–258. https://doi.org/10.1093/carcin/bgl120

Brennan K, Offiah G, McSherry EA, Hopkins AM (2010) Tight junctions: a barrier to the initiation and progression of breast cancer? J Biomed Biotechnol 2010:460607. https://doi.org/10.1155/2010/460607

Brucher B, Jamall IS (2014) Cell–cell communication in the tumor microenvironment, carcinogenesis, and anticancer treatment. Cell Physiol Biochem 34(2):213–243. https://doi.org/10.1159/000362978

Chien J, Ren Y, Qing Wang Y, Bordelon W, Thompson E, Davis R, Rayford W, Shah G (2001) Calcitonin is a prostate epithelium-derived growth stimulatory peptide. Mol Cell Endocrinol 181(1-2):69–79. https://doi.org/10.1016/s0303-7207(01)00530-5

Chlenski A, Ketels KV, Korovaitseva GI, Talamonti MS, Oyasu R, Scarpelli DG (2000) Organization and expression of the human zo-2 gene (tjp-2) in normal and neoplastic tissues. Biochim Biophys Acta 1493:319–324. https://doi.org/10.1016/s0167-4781(00)00185-8

Cording J, Berg J, Kading N, Bellmann C, Tscheik C, Westphal JK et al (2013) In tight junctions, claudins regulate the interactions between occludin, tricellulin and marvelD3, which, inversely, modulate claudin oligomerization. J Cell Sci 126:554–564. https://doi.org/10.1242/jcs.114306

Dahiya N, Becker KG, Wood WH, Zhang Y, Morin PJ (2011) Claudin-7 is frequently overexpressed in ovarian cancer and promotes invasion. PLoS One 6(7):e22119. https://doi.org/10.1371/journal.pone.0022119

Deli MA (2009) Potential use of tight junction modulators to reversibly open membranous barriers and improve drug delivery. Biochim Biophys Acta 1788(4):892–910. https://doi.org/10.1016/j.bbamem.2008.09.016

Delva E, Tucker DK, Kowalczyk AP (2009) The Desmosome. Cold Spring Harb Perspect Biol 1(2):a002543. https://doi.org/10.1101/cshperspect.a002543

Dhawan P, Singh AB, Deane NG, No Y, Shiou SR, Schmidt C, Neff J, Washington MK, Beauchamp RD (2005) Claudin-1 regulates cellular transformation and metastatic behavior in colon cancer. J Clin Invest 115(7):1765–1776. https://doi.org/10.1172/JCI24543

Dobrosotskaya I, Guy RK, James GL (1997) MAGI-1, a membrane-associated guanylate kinase with a unique arrangement of protein-protein interaction domains. J Biol Chem 272(50): 31589–31597. https://doi.org/10.1074/jbc.272.50.31589

Du H, Yang X, Fan J, Du X (2021) Claudin 6: therapeutic prospects for tumours, and mechanisms of expression and regulation (Review). Mol Med Rep 24(3):677. https://doi.org/10.3892/mmr.2021.12316. Epub 2021 Jul 23. PMID: 34296304; PMCID: PMC8335585

Dubois LG, Campanati L, Righy C, D'Andrea-Meira I, Spohr TC, Porto-Carreiro I, Pereira CM, Balça-Silva J, Kahn SA, DosSantos MF, Oliveira Mde A, Ximenes-da-Silva A, Lopes MC, Faveret E, Gasparetto EL, Moura-Neto V (2014) Gliomas and the vascular fragility of the blood brain barrier. Front Cell Neurosci 12(8):418. https://doi.org/10.3389/fncel.2014.00418

Ebnet K, Aurrand-Lions M, Kuhn A, Kiefer F, Butz S, Zander K, Meyer Brickwedde MK, Suzuki A, Imhof BA, Vestweber D (2003) The junctional adhesion molecule (JAM) family members JAM-2 and JAM-3 associate with the cell polarity protein PAR-3: a possible role for JAMs in endothelial cell polarity. J Cell Sci 116(Pt 19):3879–3891. https://doi.org/10.1242/jcs.00704

Ebnet K, Suzuki A, Ohno S, Vestweber D (2004) Junctional adhesion molecules (JAMs): more molecules with dual functions? J Cell Sci 117(Pt 1):19–29. https://doi.org/10.1242/jcs.00930

Ebnet K et al (2017) Junctional adhesion molecules (JAMs): cell adhesion receptors with pleiotropic functions in cell physiology and development. Physiol Rev 97(4):1529–1554. https://doi.org/10.1152/physrev.00004.2017

Eguchi H, Matsunaga H, Onuma S, Yoshino Y, Matsunaga T, Ikari A (2021) Down-regulation of claudin-2 expression by cyanidin-3-glucoside enhances sensitivity to anticancer drugs in the spheroid of human lung adenocarcinoma A549 cells. Int J Mol Sci 22(2):499. https://doi.org/10.3390/ijms22020499

Farquhar MG, Palade GE (1963) Junctional complexes in various epithelia. J Cell Biol 17:375–412. https://doi.org/10.1083/jcb.17.2.375

Fujiwara-Tani R, Sasaki T, Luo Y, Goto K, Kawahara I, Nishiguchi Y, Kishi S, Mori S, Ohmori H, Kondoh M, Kuniyasu H (2018) Anti-claudin-4 extracellular domain antibody enhances the antitumoral effects of chemotherapeutic and antibody drugs in colorectal cancer. Oncotarget 9(100):37367–37378. https://doi.org/10.18632/oncotarget.26427

Furuse M, Hirase T, Itoh M, Nagafuchi A, Yonemura S, Tsukita S, Tsukita S (1993) Occludin - a novel integral membrane-protein localising at tight junctions. J Cell Biol 123:1777–1788. https://doi.org/10.1083/jcb.123.6.1777

Glaunsinger BA, Weiss RS, Lee SS, Javier R (2001) Link of the unique oncogenic properties of adenovirus type 9 E4-ORF1 to a select interaction with the candidate tumor suppressor protein ZO-2. EMBO J 20:5578–5586. https://doi.org/10.1093/emboj/20.20.5578

Goodenough DA, Paul DL (2009) Gap junctions. Cold Spring Harb Perspect Biol 1(1):a002576. https://doi.org/10.1101/cshperspect.a002576

Gopalakrishnan S, Pandey N, Tamiz AP, Vere J, Carrasco R, Somerville R, Tripathi A, Ginski M, Paterson BM, Alkan SS (2009) Mechanism of action of ZOT-derived peptide AT-1002, a tight junction regulator and absorption enhancer. Int J Pharm 365(1-2):121–130. https://doi.org/10.1016/j.ijpharm.2008.08.047

Gow A, Southwood CM, Li JS, Pariali M, Riordan GP, Brodie SE, Danias J, Bronstein JM, Kachar B, Lazzarini RA (1999) CNS myelin and sertoli cell tight junction strands are absent in Osp/claudin-11 null mice. Cell 99:649–659. https://doi.org/10.1016/s0092-8674(00)81553-6

Guillemot L, Paschoud S, Pulimeno P, Foglia A, Citi S (2008) The cytoplasmic plaque of tight junctions: a scaffolding and signaling center. Biochim Biophys Acta 1778:601–613. https://doi.org/10.1016/j.bbamem.2007.09.032

Gumbiner B, Lowenkopf T, Apatira D (1991) Identification of a 160-kDa polypeptide that binds to the tight junction protein ZO-1. Proc Natl Acad Sci U S A 88:3460–3464. https://doi.org/10.1073/pnas.88.8.3460

Hamada K, Kakigawa N, Sekine S, Shitara Y, Horie T (2013) Disruption of ZO-1/claudin-4 interaction in relation to inflammatory responses in methotrexate-induced intestinal mucositis. Cancer Chemother Pharmacol 72(4):757–765. https://doi.org/10.1007/s00280-013-2238-2

Hashimoto Y, Tachibana K, Krug SM, Kunisawa J, Fromm M, Kondoh M (2019) Potential for tight junction protein-directed drug development using claudin binders and angubindin-1. Int J Mol Sci 20(16):4016. https://doi.org/10.3390/ijms20164016

Hicks DA, Galimanis CE, Webb PG, Spillman MA, Behbakht K, Neville MC, Baumgartner HK (2016) Claudin-4 activity in ovarian tumor cell apoptosis resistance and migration. BMC Cancer 16:788. https://doi.org/10.1186/s12885-016-2799-7

Hirabayashi S, Tajima M, Yao I, Nishimura W, Mori H, Hata Y (2003) JAM4, a junctional cell adhesion molecule interacting with a tight junction protein, MAGI-1. Mol Cell Biol 2003(23): 4267–4282. https://doi.org/10.1128/MCB.23.12.4267-4282.2003

Hoevel T, Macek R, Mundigl O, Swisshelm K, Kubbies M (2002) Expression and targeting of the tight junction protein CLDN1 in CLDN1-negative human breast tumor cells. J Cell Physiol 91(1):60–68. https://doi.org/10.1002/jcp.10076

Hoover KB, Liao SY, Bryant PJ (1998) Loss of the tight junction MAGUK ZO-1 in breast cancer: relationship to glandular differentiation and loss of heterozygosity. Am J Pathol 153:1767–1773. https://doi.org/10.1016/S0002-9440(10)65691-X

Ikeo K, Oshima T, Shan J, Matsui H, Tomita T, Fukui H, Watari J, Miwa H (2015) Junctional adhesion molecule-A promotes proliferation and inhibits apoptosis of gastric cancer. Hepato-Gastroenterology 62(138):540–545. PMID: 25916097

Inoko A, Itoh M, Tamura A, Matsuda M, Furuse M, Tsukita S (2003) Expression and distribution of ZO-3, a tight junction MAGUK protein, in mouse tissues. Genes Cells 8(11):837–845. https://doi.org/10.1046/j.1365-2443.2003.00681.x

Itoh M, Bissell MJ (2003) The organization of tight junctions in epithelia: implications for mammary gland biology and breast tumorigenesis. Mammary Gland Biol Neoplasia 8(4): 449–462. https://doi.org/10.1023/B:JOMG.0000017431.45314.07

Jääskeläinen A, Soini Y, Jukkola-Vuorinen A, Auvinen P, Haapasaari KM, Karihtala P (2018) High-level cytoplasmic claudin 3 expression is an independent predictor of poor survival in triple-negative breast cancer. BMC Cancer 18(1):223. https://doi.org/10.1186/s12885-018-4141-z

Jeansonne B, Lu Q, Goodenough DA, Chen YH (2003) Claudin-8 interacts with multi-PDZ domain protein 1 (MUPP1) and reduces paracellular conductance in epithelial cells. Cell Mol Biol 49(1): 13–21. PMID: 12839333

Jia H, Chai X, Li S, Wu D, Fan Z (2019) Identification of claudin-2, -6, -11 and -14 as prognostic markers in human breast carcinoma. Int J Clin Exp Pathol 12(6):2195–2204. PMID: 31934042; PMCID: PMC6949642

Joberty G, Petersen C, Gao L, Macara IG (2000) The cell-polarity protein Par6 links Par3 and atypical protein kinase C to Cdc42. Nat Cell Biol 2:531–539. https://doi.org/10.1038/35019573

Jun KH, Kim JH, Jung JH, Choi HJ, Chin HM (2014) Expression of claudin-7 and loss of claudin-18 correlate with poor prognosis in gastric cancer. Int J Surg 12(2):156–162. https://doi.org/10.1016/j.ijsu.2013.11.022

Kaihara T, Kusaka T, Nishi M, Kawamata H, Imura J, Kitajima K, Itoh-Minami R, Aoyama N, Kasuga M, Oda Y, Hattori M, Fujimori T (2003) Dedifferentiation and decreased expression of adhesion molecules, E-cadherin and ZO-1, in colorectal cancer are closely related to liver metastasis. J Exp Clin Cancer Res 22(1):117–123. PMID: 12725331

Katsuno T, Umeda K, Matsui T, Hata M, Tamura A, Itoh M, Takeuchi K, Fujimori T, Nabeshima Y, Noda T, Tsukita S, Tsukita S (2008) Deficiency of zonula occludens-1 causes embryonic lethal phenotype associated with defected yolk sac angiogenesis and apoptosis of embryonic cells. Mol Biol Cell 19(6):2465–2475. https://doi.org/10.1091/mbc.e07-12-1215

Kemphues K (2000) PARsing embryonic polarity. Cell 101:345–348. https://doi.org/10.1016/s0092-8674(00)80844-2

Kominsky SL, Argani P, Korz D, Evron E, Raman V, Garrett E, Rein A, Sauter G, Kallioniemi OP, Sukumar S (2003) Loss of the tight junction protein claudin-7 correlates with histological grade in both ductal carcinoma in situ and invasive ductal carcinoma of the breast. Oncogene 22(13): 2021–2033. https://doi.org/10.1038/sj.onc.1206199

Kominsky SL, Vali M, Korz D, Gabig TG, Weitzman SA, Argani P, Sukumar S (2004) Clostridium perfringens enterotoxin elicits rapid and specific cytolysis of breast carcinoma cells mediated through tight junction proteins claudin 3 and 4. Am J Pathol 164(5):1627–1633. https://doi.org/10.1016/S0002-9440(10)63721-2

Kornecki E, Walkowiak B, Naik UP, Ehrlich YH (1990) Activation of human platelets by a stimulatory monoclonal antibody. J Biol Chem 265:10042–10048. PMID: 2351647

Kramer F, White K, Kubbies M, Swisshelm K, Weber BH (2000) Genomic organization of claudin-1 and its assessment in hereditary and sporadic breast cancer. Hum Genet 107:249–256. https://doi.org/10.1007/s004390000375

Landers K, Samaratunga H, Teng L, Buck M, Burger MJ, Scells B, Lavin MF, Gardiner RA (2008) Identification of claudin-4 as a marker highly overexpressed in both primary and metastatic prostate cancer. Br J Cancer 99:491–501. https://doi.org/10.1038/sj.bjc.6604486

Lemmers C, Médina E, Delgrossi MH, Michel D, Arsanto JP, Le Bivic A (2002) hINADl/PATJ, a homolog of discs lost, interacts with crumbs and localizes to tight junctions in human epithelial cells. J Biol Chem 12, 277(28):25408–25415. https://doi.org/10.1074/jbc.M202196200

Leotlela PD, Wade MS, Duray PH, Rhode MJ, Brown HF, Rosenthal DT, Dissanayake SK, Earley R, Indig FE, Nickoloff BJ, Taub DD, Kallioniemi OP, Meltzer P, Morin PJ, Weeraratna AT (2007) Claudin-1 overexpression in melanoma is regulated by PKC and contributes to melanoma cell motility. Oncogene 26(26):3846–3856. https://doi.org/10.1038/sj.onc.1210155

Li J (2021) Targeting claudins in cancer: diagnosis, prognosis and therapy. Am J Cancer Res 11(7): 3406–3424. PMID: 34354852; PMCID: PMC8332862

Li WT, Jeng YM, Yang CY (2020) Claudin-18 as a marker for identifying the stomach and pancreatobiliary tract as the primary sites of metastatic adenocarcinoma. Am J Surg Pathol 44(12):1643–1648. https://doi.org/10.1097/PAS.0000000000001583

Liang J, Zhang H, Huang Y, Fan L, Li F, Li M, Yan Y, Zhang J, Li Z, Yang X (2021) A CLDN18.2-targeting bispecific T cell co-stimulatory activator for cancer immunotherapy. Cancer Manag Res 13:6977–6987. https://doi.org/10.2147/CMAR.S330637

Lioni M, Brafford P, Andl C, Rustgi A, El-Deiry W, Herlyn M, Smalley KS (2007) Dysregulation of claudin-7 leads to loss of E-cadherin expression and the increased invasion of esophageal squamous cell carcinoma cells. Am J Pathol 170(2):709–721. https://doi.org/10.2353/ajpath.2007.060343

Litkouhi B, Kwong J, Lo CM, Smedley JG 3rd., McClane BA, Aponte M, Gao Z, Sarno JL, Hinners J, Welch WR, Berkowitz RS, Mok SC, Garner EI (2007) Claudin-4 overexpression in epithelial ovarian cancer is associated with hypomethylation and is a potential target for modulation of tight junction barrier function using a C-terminal fragment of Clostridium perfringens enterotoxin. Neoplasia 9(4):304–314. https://doi.org/10.1593/neo.07118

Liu W, Li M (2020) The role of claudin-4 in the development of gastric cancer. Scand J Gastroenterol 55(9):1072–1078. https://doi.org/10.1080/00365521.2020.1795923

Liu H, Zhang Z, Zhou S, Liu X, Li G, Song B, Xu W (2021) Claudin-1/4 as directly target gene of HIF-1α can feedback regulating HIF-1α by PI3K-AKT-mTOR and impact the proliferation of esophageal squamous cell through Rho GTPase and p-JNK pathway. Cancer Gene Ther 29: 665–682. https://doi.org/10.1038/s41417-021-00328-2

Luissint AC, Lutz PG, Calderwood DA, Bourdoulous S (2008) JAM-L-mediated leukocyte adhesion to endothelial cells is regulated in cis by alpha4beta1 integrin activation. J Cell Biol 183: 1159–1173. https://doi.org/10.1083/jcb.200805061

Maher S, Leonard TW, Jacobsen J, Brayden DJ (2009) Safety and efficacy of sodium caprate in promoting oral drug absorption: from in vitro to the clinic. Adv Drug Deliv Rev 61(15): 1427–1449. https://doi.org/10.1016/j.addr.2009.09.006

Majer A, Blanchard AA, Medina S, Booth SA, Myal Y (2016) Claudin 1 expression levels affect miRNA dynamics in human basal-like breast cancer cells. DNA Cell Biol 35:328–339. https://doi.org/10.1089/dna.2015.3188

Makarova O, Roh MH, Liu CJ, Laurinec S, Margolis B (2003) Mammalian Crumbs3 is a small transmembrane protein linked to protein associated with Lin-7 (Pals1). Gene 302(1-2):21–29. https://doi.org/10.1016/s0378111902010843

Martin TA, Lane J, Ozupek H, Jiang WG (2013) Claudin-20 promotes an aggressive phenotype in human breast cancer cells. Tissue Barriers 1:e26518. https://doi.org/10.4161/tisb.26518

Martin TA, Jordan N, Davies EL, Jiang WG (2016) Metastasis to bone in human cancer is associated with loss of occludin expression. Anticancer Res 36(3):287–293. PMID: 26977027

Martìn-Padura I, Lostaglio S, Schneemann M, Williams L, Romano M, Fruscella P, Panzeri C, Stoppacciaro A, Ruco L, Villa A, Simmons D, Dejana E (1998) Junctional adhesion molecule, a novel member of the immunoglobulin superfamily that distributes at intercellular junctions and modulates monocyte transmigration. J Cell Biol 142(1):117–127. https://doi.org/10.1083/jcb.142.1.117

Mašindová I, Šoltýsová A, Varga L, Mátyás P, Ficek A, Hučková M, Sůrová M, Šafka-Brožková D, Anwar S, Bene J, Straka S, Janicsek I, Ahmed ZM, Seeman P, Melegh B, Profant M, Klimeš I, Riazuddin S, Kádasi L, Gašperíková D (2015) MARVELD2 (DFNB49) mutations in the hearing impaired central European Roma population - prevalence, clinical impact and the common origin. PLoS ONE 10(4):e0124232. https://doi.org/10.1371/journal.pone.0124232

Mauro L, Bartucci M, Morelli C, Ando S, Surmacz E (2001) IGF-I receptor-induced cell-cell adhesion of MCF-7 breast cancer cells requires the expression of junction protein ZO-1. J Biol Chem 276:39892–39897. https://doi.org/10.1074/jbc.M106673200

Miyamoto K, Kusumi T, Sato F, Kawasaki H, Shibata S, Ohashi M, Hakamada K, Sasaki M, Kijima H (2008) Decreased expression of claudin-1 is correlated with recurrence status in esophageal squamous cell carcinoma. Biomed Res 29(2):71–76. https://doi.org/10.2220/biomedres.29.71. PMID: 18480547

Moiseev M, Khutoryanskiy S (2019) Penetration enhancers in ocular drug delivery. Pharmaceutics 11:321. https://doi.org/10.3390/pharmaceutics11070321

Morampudi V, Graef FA, Stahl M, Dalwadi U, Conlin VS, Huang T, Vallance BA, Yu HB, Jacobson K (2016) Tricellular tight junction protein tricellulin is targeted by the enteropathogenic Escherichia coli effector EspG1, leading to epithelial barrier disruption. Infect Immun 85: e00700–e00716

Mori H, Hirabayashi S, Shirasawa M, Sugimura H, Hata Y (2004) JAM4 enhances hepatocyte growth factor-mediated branching and scattering of Madin-Darby canine kidney cells. Genes Cells 9:811–819. https://doi.org/10.1111/j.1365-2443.2004.00765.x

Morin PJ (2005) Claudin proteins in human cancer: promising new targets for diagnosis and therapy. Cancer Res 65(21):9603–9606. https://doi.org/10.1158/0008-5472.CAN-05-2782

Morita K, Sasaki H, Furuse M, Tsukita S (1999) Endothelial claudin: claudin-5/TMVCF constitutes tight junction strands in endothelial cells. J Cell Biol 147:185–194. https://doi.org/10.1083/jcb.147.1.185

Morrison PWJ, Khutoryanskiy VV (2014) Enhancement in corneal permeability of riboflavin using calcium sequestering compounds. Int J Pharm 472:56–64. https://doi.org/10.1016/j.ijpharm.2014.06.007

Naik MU, Naik TU, Suckow AT, Duncan MK, Naik UP (2008) Attenuation of junctional adhesion molecule-A is a contributing factor for breast cancer cell invasion. Cancer Res 68(7): 2194–2203. https://doi.org/10.1158/0008-5472.CAN-07-3057

Nissi R, Talvensaari-Mattila A, Kuvaja P, Pääkkö P, Soini Y, Santala M (2015) Claudin-5 is associated with elevated TATI and CA125 levels in mucinous ovarian borderline tumors. Anticancer Res 35(2):973–976. PMID: 25667483

Oda Y, Otani T, Ikenouchi J, Furuse M (2014) Tricellulin regulates junctional tension of epithelial cells at tricellular contacts through Cdc42. J Cell Sci 127(19):4201–4212. https://doi.org/10.1242/jcs.150607

Offner S, Hekele A, Teichmann U, Weinberger S, Gross S, Kufer P, Itin C, Baeuerle PA, Kohleisen B (2005) Epithelial tight junction proteins as potential antibody targets for pancarcinoma therapy. Cancer Immunol Immunother 54(5):431–445. https://doi.org/10.1007/s00262-004-0613-x

Osanai M, Murata M, Nishikiori N, Chiba H, Kojima T, Sawada N (2006) Epigenetic silencing of occludin promotes tumorigenic and metastatic properties of cancer cells via modulations of unique sets of apoptosis-associated genes. Cancer Res 66(18):9125–9133. https://doi.org/10.1158/0008-5472.CAN-06-1864

Osanai M, Murata M, Nishikiori N, Chiba H, Kojima T, Sawada N (2007) Occludin-mediated premature senescence is a fail-safe mechanism against tumorigenesis in breast carcinoma cells. Cancer Sci 98(7):1027–1034. https://doi.org/10.1111/j.1349-7006.2007.00494.x

Ozsoy Y, Gungor S, Cevher E (2009) Nasal delivery of high molecular weight drugs. Molecules 14(9):3754–3779. https://doi.org/10.3390/molecules14093754

Park JY, Park KH, Oh TY, Hong SP, Jeon TJ, Kim CH, Park SW, Chung JB, Song SY, Bang S (2007) Up-regulated claudin 7 expression in intestinal-type gastric carcinoma. Oncol Rep 18 (2):377–382. PMID: 17611659

Qin Z, Fang D, Fang Y (2017) Low expression of occludin: a predictor of poor prognosis in esophageal squamous cell carcinoma. Int J Clin Exp Pathol 10(7):7451–7459. PMID: 31966588; PMCID: PMC6965274

Qiu RG, Abo A, Steven Martin G (2000) A human homolog of the C. elegans polarity determinant Par-6 links Rac and Cdc42 to PKCzeta signaling and cell transformation. Curr Biol 10:697–707. https://doi.org/10.1016/s0960-9822(00)00535-2

Reeh M, Bockhorn M, Görgens D, Vieth M, Hoffmann T, Simon R, Izbicki JR, Sauter G, Schumacher U, Anders M (2013) Presence of the coxsackievirus and adenovirus receptor (CAR) in human neoplasms: a multitumour array analysis. Br J Cancer 109(7):1848–1858. https://doi.org/10.1038/bjc.2013.509

Resnick MB, Konkin T, Routhier J, Sabo E, Pricolo VE (2005) Claudin-1 is a strong prognostic indicator in stage II colonic cancer: a tissue microarray study. Mod Pathol 18(4):511–518. https://doi.org/10.1038/modpathol.3800301

Roh MH, Liu CJ, Laurinec S, Margolis B (2002) The carboxyl terminus of zona occludens-3 binds and recruits a mammalian homologue of discs lost to tight junctions. J Biol Chem 277(30): 27501–27509. https://doi.org/10.1074/jbc.M201177200

Saitou M, Furuse M, Sasaki H, Schulzke JD, Fromm M, Takano H, Noda T, Tsukita S (2000) Complex phenotype of mice lacking occludin, a component of tight junction strands. Mol Biol Cell 11:4131–4142. https://doi.org/10.1091/mbc.11.12.4131

Salvador E, Burek M, Förster CY (2015) Stretch and/or oxygen glucose deprivation (OGD) in an in vitro traumatic brain injury (TBI) model induces calcium alteration and inflammatory cascade. Front Cell Neurosci 9:323. https://doi.org/10.3389/fncel.2015.00323

Salvador E, Burek M, Forster CY (2016) Tight junctions and the tumor microenvironment. Curr Pathobiol Rep 4:135–145. https://doi.org/10.1007/s40139-016-0106-6

Seo KW, Kwon YK, Kim BH, Kim CI, Chang HS, Choe MS, Park CH (2010) Correlation between claudins expression and prognostic factors in prostate cancer. Korean J Urol 51(4):239–244. https://doi.org/10.4111/kju.2010.51.4.239

Shang X, Lin X, Alvarez E, Manorek G, Howell SB (2012) Tight junction proteins claudin-3 and claudin-4 control tumor growth and metastases. Neoplasia 14(10):974–985. https://doi.org/10.1593/neo.12942

Shen Z, Cao B, Lin L, Zhou C, Ye D, Qiu S, Li Q, Cui X (2017) The clinical signification of claudin-11 promoter hypermethylation for laryngeal squamous cell carcinoma. Med Sci Monit 23:3635–3640. https://doi.org/10.12659/msm.904751

Shin DY, Lee WS, Jung JH, Hong SH, Park C, Kim HJ, Kim GY, Hwang HJ, Kim GS, Jung JM, Ryu CH, Shin SC, Hong SC, Choi YH (2013) Flavonoids from Orostachys japonicus a. berger inhibit the invasion of lncap prostate carcinoma cells by inactivating akt and modulating tight junctions. Int J Mol Sci 14(9):18407–18420. https://doi.org/10.3390/ijms140918407

Sobel G, Páska C, Szabó I, Kiss A, Kádár A, Schaff Z (2005) Increased expression of claudins in cervical squamous intraepithelial neoplasia and invasive carcinoma. Hum Pathol 36(2): 162–169. https://doi.org/10.1016/j.humpath.2004.12.001

Soini Y (2011) Claudins in lung diseases. Respir Res 12(1):70. https://doi.org/10.1186/1465-9921-12-70

Soini Y (2012) Tight junctions in lung cancer and lung metastasis: a review. International Journal of. Clin Exp Pathol 5(2):126–136. PMID: 22400072; PMCID: PMC3294225

Stadler CR, Bähr-Mahmud H, Plum LM, Schmoldt K, Kölsch AC, Türeci Ö, Sahin U (2015) Characterization of the first-in-class T-cell-engaging bispecific single-chain antibody for targeted immunotherapy of solid tumors expressing the oncofetal protein claudin 6. Onco Targets Ther 5(3):e1091555. https://doi.org/10.1080/2162402X.2015.1091555

Tabariès S, Dupuy F, Dong Z, Monast A, Annis MG, Spicer J, Ferri LE, Omeroglu A, Basik M, Amir E, Clemons M, Siegel PM (2012) Claudin-2 promotes breast cancer liver metastasis by facilitating tumor cell interactions with hepatocytes. Mol Cell Biol 32(15):2979–2991. https://doi.org/10.1128/MCB.00299-12

Tabariès S, Annis MG, Lazaris A, Petrillo SK, Huxham J, Abdellatif A, Palmieri V, Chabot J, Johnson RM, Van Laere S, Verhoef C, Hachem Y, Yumeen S, Meti N, Omeroglu A, Altinel G, Gao ZH, Yu ASL, Grünhagen DJ, Vermeulen P, Metrakos P, Siegel PM (2021) Claudin-2 promotes colorectal cancer liver metastasis and is a biomarker of the replacement type growth pattern. Commun Biol 4(1):657. https://doi.org/10.1038/s42003-021-02189-9

Tachibana K, Kondoh M (2021) A method to prepare a bioprobe for regulatory science of the drug delivery system to the brain: an angulin binder to modulate tricellular tight junction-seal. Methods Mol Biol 2367:291–304. https://doi.org/10.1007/7651_2020_317

Tachibana K, Iwashita Y, Wakayama E, Nishino I, Nishikaji T, Kondoh M (2020) Tight junction modulating bioprobes for drug delivery system to the brain: a review. Pharmaceutics 12(12): 1236. https://doi.org/10.3390/pharmaceutics12121236

Tobioka H, Tokunaga Y, Isomura H, Kokai Y, Yamaguchi J, Sawada N (2004) Expression of occludin, a tight-junction-associated protein, in human lung carcinomas. Virchows Arch 445(5): 472–476. https://doi.org/10.1007/s00428-004-1054-9

Törzsök P, Riesz P, Kenessey I, Székely E, Somorácz A, Nyirády P, Romics I, Schaff Z, Lotz G, Kiss A (2011) Claudins and ki-67: potential markers to differentiate low- and high-grade transitional cell carcinomas of the urinary bladder. J Histochem Cytochem 59(11):1022–1030. https://doi.org/10.1369/0022155411424606

Uchida H, Kondoh M, Hanada T, Takahashi A, Hamakubo T, Yagi K (2010) A claudin-4 modulator enhances the mucosal absorption of a biologically active peptide. Biochem Pharmacol 79(10): 1437–1444. https://doi.org/10.1016/j.bcp.2010.01.010

Ungureanu BS, Lungulescu CV, Pirici D, Turcu-Stiolica A, Gheonea DI, Sacerdotianu VM, Liliac IM, Moraru E, Bende F, Saftoiu A (2021) Clinicopathologic relevance of claudin 18.2 expression in gastric cancer: a meta-analysis. Front Oncol 11:643872. https://doi.org/10.3389/fonc.2021.643872

Usami Y, Chiba H, Nakayama F, Ueda J, Matsuda Y, Sawada N, Komori T, Ito A, Yokozaki H (2006) Reduced expression of claudin-7 correlates with invasion and metastasis in squamous cell carcinoma of the esophagus. Hum Pathol 37(5):569–577. https://doi.org/10.1016/j.humpath.2005.12.018

Wada M, Tamura A, Takahashi N, Tsukita S (2013) Loss of claudins 2 and 15 from mice causes defects in paracellular Na? Flow and nutrient transport in gut and leads to death from malnutrition. Gastroenterology 144(2):369–380. https://doi.org/10.1053/j.gastro.2012.10.035

Wang ZL, Mandell KJ, Parkos CA, Mrsny RJ, Nusrat A (2005) The second loop of occludin is required for suppression of Raf1-induced tumor growth. Oncogene 24(27):4412–4420. https://doi.org/10.1038/sj.onc.1208634

Wang Q, Zhang Y, Zhang T, Han ZG, Shan L (2015) Low claudin-6 expression correlates with poor prognosis in patients with non-small cell lung cancer. Onco Targets Ther 8:1971–1977. https://doi.org/10.2147/OTT.S85478

Xu X, You K, Wu B (2019) Zonula occludens-1 associated nucleic acid binding protein plays an invasion-promoting role in bladder cancer. Neoplasma 66(3):405–419. https://doi.org/10.4149/neo_2018_180725N530. Epub 2019 Feb 14. PMID: 30784285

Yang G, Jian L, Chen Q (2021) Comprehensive analysis of expression and prognostic value of the claudin family in human breast cancer. Aging 13(6):8777–8796. https://doi.org/10.18632/aging.202687

Zeisel MB, Dhawan P, Baumert TF (2019) Tight junction proteins in gastrointestinal and liver disease. Gut 68(3):547–561. https://doi.org/10.1136/gutjnl-2018-316906. Epub 2018 Oct 8. PMID: 30297438; PMCID: PMC6453741

Zhang X, Wang H, Li Q, Li T (2018) CLDN2 inhibits the metastasis of osteosarcoma cells via down-regulating the afadin/ERK signaling pathway. Cancer Cell Int 18:160. https://doi.org/10.1186/s12935-018-0662-4

Zhu L, Han J, Li L, Wang Y, Li Y, Zhang S (2019) Claudin family participates in the pathogenesis of inflammatory bowel diseases and colitis-associated colorectal cancer. Front Immunol 10: 1441. https://doi.org/10.3389/fimmu.2019.01441

Zinner B, Gyöngyösi B, Babarczi E, Kiss A, Sobel G (2013) Claudin 1 expression characterizes human uterine cervical reserve cells. J Histochem Cytochem 61(12):880–888. https://doi.org/10.1369/0022155413501324

The manufacturer's authorised representative in the EU is Springer Nature Customer Service Centre GmbH, Europaplatz 3, 69115 Heidelberg, Germany. If you have any concerns regarding our products, please contact ProductSafety@springernature.com

Printed and bound by CPI Group (UK) Ltd, Croydon, CR0 4YY
30/09/2026
02225319-0001